Active Sensor Planning for Multiview Vision Tasks

Shengyong Chen · Youfu Li ·
Jianwei Zhang · Wanliang Wang

Active Sensor Planning
for Multiview Vision Tasks

Springer

Shengyong Chen
College of Information Engineering
Zhejiang University of Technology
Hangzhou 310014, PR China
sy@ieee.org

Jianwei Zhang
AB TAMS, Department of Informatics
University of Hamburg
Vogt-Koelln-Strasse 30, D-22527
Hamburg, Germany
zhang@informatik.uni-hamburg.de

Y. F. Li
Dept of Manufacturing Engineering
and Engineering Management
City University of Hong Kong
83 Tat Chee Avenue, Kowloon
Hong Kong
meyfli@cityu.edu.hk

Wanliang Wang
College of Software Engineering
Zhejiang University of Technology
Hangzhou 310014
PR China

ISBN: 978-3-540-77071-8   e-ISBN: 978-3-540-77072-5

Library of Congress Control Number: 2007940478

© 2008 Springer-Verlag Berlin Heidelberg

Cover design: WMXDesign GmbH

Printed on acid-free paper

9 8 7 6 5 4 3 2 1

springer.com

Preface

The problem of active sensor planning was firstly addressed about 20 years ago and attracted many people after then. Recently, active sensing becomes even more important than ever since a number of advanced robots are available now and many tasks require to act actively for obtaining 3D visual information from different aspects. Just like human beings, it's unimaginable if without active vision even only in one minute. Being active, the active sensor planner is able to manipulate sensing parameters in a controlled manner and performs active behaviors, such as active sensing, active placement, active calibration, active model construction, active illumination, etc. Active vision perception is an essential means of fulfilling such vision tasks that need take intentional actions, e.g. entire reconstruction of an unknown object or dimensional inspection of an industrial workpiece.

The intentional actions introduce active or purposive behaviors. The vision system (the observer) takes intentional actions according to its mind, the mind such as going to a specific location and obtaining the useful information of the target, in an uncertain environment and conditions. It has a strategic plan to finish a certain vision task, such as navigating through an unfamiliar environment or modeling of an unknown object. It is capable of executing the plan despite the presence of unanticipated objects and noisy sensors.

A multi-view strategy is often required for seeing object features from optimal placements since vision sensors have limited field of view and can only "see" a portion of a scene from a single viewpoint. This means that the performance of a vision sensor depends heavily both on the type and number of sensors and on the configuration of each sensor. What important is that the sensor is active. Compared with the typical passive vision where it is limited to what is offered by the preset visual parameters and environmental conditions, the active planner can instead determine how to view by utilizing its capability to change its visual parameters according to the scene for a specific task at any time.

From this idea, many problems have to be considered in constructing an active perception system and these important problems lead our motivation of the research on active sensor and sensing techniques. For many practical vision tasks, because, it is very necessary to develop a multiview plan of control strategy, and these viewpoints can be decided either offline or in run-time.

The purpose of this book is to introduce the challenging problems and propose some possible solutions. The main topics addressed are from both theoretical and technical aspects, including sensing activity, configuration, calibration, sensor modeling, sensing constraints, sensing evaluation, viewpoint decision, sensor placement graph, model based planning, path planning, planning for unknown environment, incremental 3D model construction, measurement, and surface analysis.

With our acknowledgements, the research work in this book was supported by the Research Grants Council of Hong Kong, the Natural Science Foundation of China, and the Alexander von Humboldt Foundation of Germany. Several colleagues, Z. Liu, B. He, B. Zhang, H. Zhang, and T. Tetsis, have contributed in part with the research work, experiments, and writing.

Shengyong Chen
Zhejiang University of Technology (China)

Contents

Chapter 1
Introduction

Active sensor planning is an important means for fulfilling vision tasks that require intentional actions, e.g. complete reconstruction of an unknown object or dimensional inspection of a workpiece. Constraint analysis, active sensor placement, active sensor configuration, and three-dimensional (3D) data acquisition are the essential steps in developing such active vision systems. This chapter presents the general motivations, ideas for solutions, and potential applications of active sensor planning for multiview vision tasks.

1.1 Motivations

The *intentional actions* in active sensor planning for visual perception introduce *active behaviors* or *purposeful behaviors*. The vision agent (the observer) takes intentional actions according to its set goal such as going to a specific location or obtaining the full information on an object, in the current environment and its own state. A strategic plan is needed to finish a vision task, such as navigating through an office environment or modeling an unknown object. In this way, the plan can be executed successfully despite the presence of unanticipated objects and noisy sensors.

Therefore there are four aspects that need to be studied in developing an active observer, i.e. the *sensor itself* (its type and measurement principle), the *sensor state* (its configuration and parameters), the *observer state* (its pose), and *the planner* for scene interpretation and action decision. Although many other things have to be considered in constructing an active perception system, these important issues lead to the research on active visual perception and investigations on visual sensing, system reconfiguration, automatic sensor planning, and interpretation and decision.

1.1.1 The Tasks

A critical problem in modern robotics is to endow the observer with a strategic plan to finish certain tasks. The multi-view strategy is an important means of taking active actions in visual perception, by which the vision sensor is purposefully placed at several positions to observe a target.

Sensor planning which determines the pose and configuration for the visual sensor thus plays an important role in active vision perception, not only because a 3D sensor has a limited field of view and can only see a portion of a scene from a single viewpoint, but also because a global description of objects often cannot be reconstructed from only one viewpoint due to occlusion. Multiple viewpoints have to be planned for many vision tasks to make the entire object (or all the features of interest) visible strategically.

For tasks of observing unknown objects or environments, the viewpoints have to be decided in run-time because there is no prior information about the targets. Furthermore, in an inaccessible environment, the vision agent has to be able to take intentional actions automatically. The fundamental objective of sensor placement in such tasks is to increase knowledge about the unseen portions of the viewing volume while satisfying all the placement constraints such as in-focus, field-of-view, occlusion, collision, etc. An optimal viewpoint planning strategy determines each subsequent vantage point and offers the obvious benefit of reducing and eliminating the labor required to acquire an object's surface geometry. A system without planning may need as many as seventy range images for recovering a 3D model with normal complexity, with significant overlap between them. It is possible to reduce the number of sensing operations to less than ten times with a proper sensor planning strategy. Furthermore, it also makes it possible to create a more accurate and complete model by utilizing a physics-based model of the vision sensor and its placement strategy.

For model-based tasks, especially for industrial inspections, the placements of the sensor need to be determined and optimized before robot operations. Generally in these tasks, the sensor planning is to find a set of admissible viewpoints in the acceptable space, which satisfy all of the sensor placement constraints and can finish the vision task well. In most of the related works, the constraints in sensor placement are expressed as a cost function with the aim to achieve the minimum cost. However, previously the evaluation of a viewpoint has normally been achieved by direct computation. Such an approach is usually formulated for a particular application and is therefore difficult to be applied to general tasks. For a multi-view sensing strategy, global optimization is desired but was rarely considered in the past.

In an active vision system, the visual sensor has to be moved frequently for purposeful visual perception. Since the targets may vary in size and distance and the task requirements may also change in observing different objects or features, a structure-fixed vision sensor is usually insufficient. For a structured light vision sensor, the camera needs to be able to "see" just the scene illuminated by the projector. Therefore the configuration of a vision setup often needs to be changed to reflect the constraints in different views and achieve optimal acquisition performance. A reconfigurable sensor can change its structural parameters to adapt itself to the scene to obtain maximum 3D information from the target.

In practical applications, in order to reconstruct an object with high accuracy, it is essential that the vision sensor be carefully calibrated. Traditional calibration methods are mainly for static uses in which a calibration target with specially designed features needs to be placed at precisely known locations. However, when the sensor is reconfigured, it must be re-calibrated again. To avoid the tedious and laborious procedures in such traditional calibrations, a self-recalibration method is needed to perform the task automatically so that 3D reconstruction can follow immediately without a calibration apparatus and any manual interference.

Finally, 3D reconstruction is either an ultimate goal or a means to the goal in 3D computer vision. For some tasks, such as reverse engineering and constructing environments for virtual reality, the 3D reconstruction of the target is the goal of the vision perception. A calibrated vision sensor which applies the 3D sensing techniques is used to measure the object surfaces in the scene. Then the obtained local models are globally integrated into a complete model for describing the target shape. For some other vision tasks, such as object recognition and industrial inspection, the 3D reconstruction is an important means to achieve the goal. In such a case, the 3D measurement is performed at every step for making a decision or drawing a conclusion about the target.

1.1.2 From a Biological View

I move, therefore I see. (Hamada 1992)
Active sensor planning now plays a most important role in practical vision systems because generally a global description of objects cannot be reconstructed from only one viewpoint due to occlusion or limited Field Of View (FOV). For example, in the case of object modeling tasks, because there is no prior information about the objects or environments, it is obviously very necessary to develop a multi-view plan of a controlling strategy, and these views can be decided either in run-time or off-line.

To illustrate the strong relationship between active perception and multi-view sensor planning, we may begin the explanation with a look of human behaviors. In humans, the operations and informational contents of the global state variable, which are sensations, images, feelings, thoughts and beliefs, constitute the experience of causation. In the field of neuro-dynamics and causality, Freeman (1999a) used circular causality to explain neural pattern formation by self-organizing dynamics. It explained how stimuli cause consciousness by referring to causality. An aspect of intentional action is causality, which we extrapolate to material objects in the world. Thus causality is a property of mind, not matter.

In the biological view (Fig. 1.1), in a passive information processing system a stimulus input gives information (Freeman 1999b), which is transduced by receptors into trains of impulses that signify the features of an object. Symbols are processed according to rules for learning and association and are then bound into a representation, which is stored, retrieved and matched with new incoming representations. In active systems perception begins with the emergence of a goal that is implemented by the search for information. The only input accepted is that which is consistent with the goal and anticipated as a consequence of the searching actions. The key component to be modeled in brains provides the dynamics that constructs goals and the adaptive actions by which they are achieved.

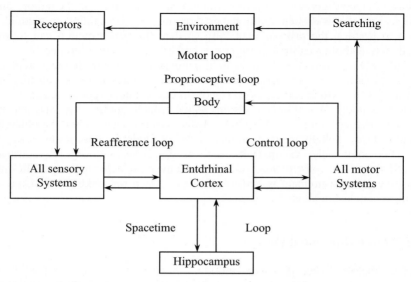

Fig. 1.1. The proprioceptive loop of human beings (Freeman 1999b)

1.1.3 The Problems and Goals

Many applications in robotics involve a good knowledge of the robot environment. 3D machine vision is the technology which allows computers to measure the three-dimensional shape of objects or environments, without resorting to physically probing their surfaces. The object/environment model is constructed in three stages. First, apply a computer vision technique called "shape from X" (e.g. shape from stereo) to determine the shapes of the objects visible in each image. The second stage is to integrate these image-based shape models into a single, complete shape model of the entire scene. Third, finally the shape model is rendered with the same color of the real object.

In developing such a technique, sensor planning is a critical issue since a typical 3D sensor can only sample a portion of an object at a single viewpoint. Using a vision sensor to sample all of the visible surfaces of any but the most trivial of objects, however, requires that multiple 3D images be taken from different vantage points and integrated, i.e. merged to form a complete model (Fig. 1.2). An optimal viewpoint planning strategy (or next best view – NBV algorithm) determines each subsequent vantage point and offers the obvious benefit of reducing and eliminating the labor required to acquire an object's surface geometry.

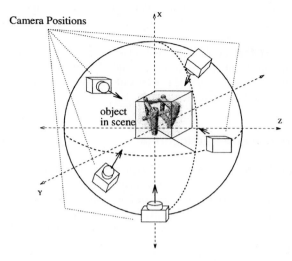

Fig. 1.2. A sequence of sensor pose placements for object modeling (Banta and Abidi 1996), p1→p2→...→pn

On the other hand, the performance of the vision perception of a robot, and thus the quality of knowledge learnt, can be significantly affected by the properties of illumination such as intensity and color. "Where to place the illuminants and how to set the illuminant parameters" for improving the process of vision tasks becomes an increasingly important problem that needs be solved. Drawing on the wide body of knowledge in radiometry and photometry will prove useful. For this purpose, this book also focuses attention on the topic of illumination planning in the robot vision system. Illumination planning can be considered as a part of the sensor planning problem for vision tasks. This book presents a study on obtaining the optimal illumination condition (or most comfortable condition) for vision perception. The "comfortable" condition for a robot eye is defined as: the image has a high signal-to-noise ratio and high contrast, is within the linear dynamic range of the vision sensor, and reflects the natural properties of the concerned object. "Discomfort" may occur if any of these criteria are not met because some scene information may not be recorded. This book also proposes appropriate methods to optimize the optical parameters of the luminaire and the sensor (including source radiant intensity, sensor aperture and focus length) and the pose parameters of the luminaire, with emphasis on controlling the intensity and avoiding glare.

The proposed strategy to implement placements of vision sensors and light sources requires an eye-and-hand setup allowing the sensors (a pair of stereo cameras or a structure of projector+camera) to be moving around and looking/shining at an object from different viewpoints. The sensor or luminaire is mounted on the end-effector of a robot to achieve arbitrary spatial position and orientation. The purpose of moving the camera is to arrive at viewing poses, such that required details of the unknown object can be acquired by the vision system for reconstruction of a 3D model. This book ignores the problems of the relative orientation between cameras and manipulator coordinate systems, and how to control the sensor movement.

Another goal of this book is to demonstrate the interdependence between a solution to the sensor planning problem and the other stages of vision image process, as well as the necessity and benefits of utilizing a model of the sensor when determining sensor setting and viewpoint.

1.1.4 Significance and Applications

The techniques described in this book may have outstanding significance in the many applications of computer vision. Using artificial vision for 3D object reconstruction and modeling is the technology which allows computers to obtain the three-dimensional shape of objects, without resorting to physically probing their surfaces. This is best suited for tasks where non-contact nature, a fast measurement rate and cost are of primary concern, especially for such applications as:

- medical applications,
- archeology,
- quality ensurance,
- reverse engineering,
- rapid product design,
- robotics, etc.

The technology of active sensor planning has its significance in both model based and non-model based applications of computer vision. Typical non-model based applications include:

- 3D object reconstruction and modeling,
- target searching,
- scene exploration,
- autonomous navigation, etc.

Model-based applications, where the object's geometry and a rough estimate of its pose are known, are widely used in:

- product assembly/disassembly,
- feature detection,
- inspection,
- object recognition,
- searching,
- dimensional measurement,
- surveillance,
- target tracking,
- monitoring, etc.

1.2 Objectives and Solutions

The general aim of this book is to introduce the ideas for possible solutions for the above-motivated problems in an active visual system. The objectives of the research include the following:

- To introduce the guideline for setting up typical active vision systems and applying it to 3D visual perception;
- To develop methods of active reconfiguration for purposive visual sensing;
- To investigate methodologies of automatic sensor planning for industrial applications;
- To propose strategies for sensor planning incorporation with illumination planning;
- To find solutions for exploring the 3D structure of an unknown target.

Among all of the above, this book places its emphasis on the last three issues. In the study of sensor planning, previous approaches mainly focused on the modeling of sensor constraints and calculating a "good" viewpoint for observing one or several features on the object. Little consideration was given to the overall efficiency of a generated plan with a sequence of viewpoints. In model-based vision tasks, researchers have made efforts to find an admissible domain of viewpoints to place the sensor to look at one or several object features. However, this method is difficult to apply in a multi-feature-multi-viewpoint problem as it cannot determine the minimum number of viewpoints and their relative distribution.

In non-model-based vision tasks, previous research efforts often concentrated on finding the best next views by volumetric analysis or occlusion as a guide. However, since no information about the unknown target exists, it is actually impossible to give the true best next view. It exists but can only be determined after the complete model has been obtained. Therefore a critical problem is still not well solved: the global optimization of sensor planning. When multiple features need to be observed and multiple viewpoints need to be planned, the minimum number of viewpoints needs to be determined. To achieve high efficiency and quality, the optimal spatial distribution of the viewpoints should be determined too. These are also related to the sensor configuration and environmental constraints. Furthermore, to make it flexible in practical applications, we need to deal with arbitrary object models without assumptions on the object features.

In this book, ideas are presented to solve the relevant issues in active sensing problems. Novel methodologies are developed in sensor configuration, 3D reconstruction, sensor placement, and viewpoint planning for multiview vision tasks.

In setting up the active vision system for the study, both traditional stereo cameras and coded structured light systems are investigated. The coded light approach can be adopted for the digital projector to generate binary patterns with light and dark stripes which are switchable during the operation. This method features high accuracy and reliability.

The sensor planning presented in this book is an effective strategy to generate a sequence of viewing poses and corresponding sensor configurations for optimal completion of a multiview vision task. Methods are proposed to solve the problems for both model-based and non-model-based vision tasks. For model-based applications, the method involves the determination of the optimal sensor placements and the shortest path through the viewpoints for automatic generation of a perception plan. A topology of the viewpoints is achieved by a genetic algorithm in which a min-max criterion is used for evaluation. The shortest path is also determined by graph theory. The sensing plan generated by the proposed methods leads to global optimization. For non-model-based applications, the method involves the decision of the exploration direction and the determination of the best next view and the corresponding sensor settings. Some cues are proposed to predict the unknown portion of an object or environment and the next best viewpoint is determined by the expected surface. The viewpoint determined in such a way is predictably best. Information Entropy Based Planning, uncertainty-driven planning, and self-termination conditions are also discussed.

Numerical simulations and practical experiments are conducted to implement the proposed methods for the active sensing in the multi-view vision task. The implementation results obtained in these initial experiments are only intended for showing the validity of proposed methods and the feasibility for practical applications. Using the active visual perception strategy, 3D reconstruction can be achieved without the constraints on the system configuration parameters. This allows optimal system configuration to be employed to adaptively sense an environment. The proposed methods will give the active vision system the adaptability needed in many practical applications.

1.3 Book Structure

This book is organized as follows.

- *Chapter 2* presents the sensing fundamentals, measurement principles, and 3D reconstruction methods for active visual sensing. These will be used in the next chapters in formulating the methods of sensor planning. It also describes the methods for dynamic reconfiguration and recalibration of a stripe light vision system to overcome practical scene challenge.
- *Chapter 3* summarizes the relevant works on 3D sensing techniques and introduces the active 3D visual sensing systems developed in the community. Both stereo sensors and structured light systems are mainly considered in this book, although extensions of other types of visual sensors such as laser scanners are straightforward.
- *Chapter 4* presents the sensor models, summarizes the previous approaches to the sensor planning problem, formulates sensor placement constraints, and proposes the criteria for plan evaluation. The method for the model-based sensor placement should meet both the optimal sensor placements and the shortest path

through these viewpoints. The plan for such sensor placements is evaluated based on the fulfillment of three conditions: low order, high precision, and satisfying all constraints.

- *Chapter 5* presents a method for automatic sensor placement for model-based robot vision. The task involves determination of the optimal sensor placements and a shortest path through these viewpoints. During the sensor planning, object features are resampled as individual points attached to surface normals. The optimal sensor placement graph is achieved by a genetic algorithm in which a min-max criterion is used for the evaluation. A shortest path is determined by graph theories. A viewpoint planner is developed to generate the sensor placement plan.

- *Chapter 6* presents a sensing strategy for determining the probing points for achieving efficient measurement and reconstruction of freeform surfaces. The B-spline model is adopted for modeling the freeform surface. In order to obtain reliable parameter estimation for the B-spline model, we analyze the uncertainty of the model and use the statistical analysis of the Fisher information matrix to optimize the locations of the probing points needed in the measurements.

- *Chapter 7* presents the issues regarding sensor planning for incrementally building a complete model of an unknown object or environment by an active visual system. It firstly lists some typical approaches to sensor planning for model construction, including the multi-view strategy and existing contributions. The standard procedure for modeling of unknown targets is provided. A self-termination judgment method is suggested based on Gauss' Theorem by checking the variations of the surface integrals between two successive viewpoints so that the system knows when the target model is complete and it is necessary to stop the modeling procedure.

- *Chapter 8* presents an information entropy-based sensor planning approach for the reconstruction of freeform surfaces of 3D objects. In the framework of Bayesian statistics, it proposes an improved Bayesian information criterion (BIC) for determining the B-spline model complexity. Then, the uncertainty of the model is analyzed using entropy as the measurement. Based on this analysis, the information gain for each cross section curve is predicted for the next measurement. After predicting the information gain of each curve, we can obtain the information change for all the B-spline models. This information gain is them mapped into the view space. The viewpoint that contains maximal information gain about the object is selected as the Next Best View.

- *Chapter 9* also deals with the sensor placement problem, but for the tasks of non-model based object modeling. The method involves the decision of the exploration direction and the determination of the best next view and the corresponding sensor settings. The trend surface is proposed as the cue to predict the unknown portion of an object.

- *Chapter 10* presents some strategies of adaptive illumination control for robot vision to achieve the best scene interpretation. It investigates how to obtain the most comfortable illumination conditions for a vision sensor. Strategies are proposed to optimize the pose and optical parameters of the luminaire and the sensor, with emphasis on controlling the image brightness.

Chapter 2
Active Vision Sensors

This chapter presents the sensing fundamentals, measurement principles, and 3D reconstruction methods for active visual sensing. An idea of sensor reconfiguration and recalibration is also described which endows a robot with the ability of actively changing its sensing parameters according to practical scenes, targets, and purposes. These will be used in the next chapters in formulating the methods of sensor reconfiguration and sensor planning.

2.1 3D Visual Sensing by Machine Vision

Similar to human perception, machine vision perception is one of the most important ways for acquiring knowledge of the environment. The recovery of the 3D geometric information of the real world is a challenging problem in computer vision research. Active research in the field in the last 30 years has produced a huge variety of techniques for 3D sensing. In robotic applications, the 3D vision technology allows computers to measure the three-dimensional shape of objects or environments, without resorting to physically probing their surfaces.

2.1.1 Passive Visual Sensing

One class of visual sensing methods is called passive visual sensing where no other device besides cameras is required. These methods were usually developed at the early stage of computer vision research. By passive, no energy is emitted for the sensing purpose and the images are the only input data. The sensing techniques were often supposed to reflect the way that human eyes work. The limited equipment cost constitutes a competitive advantage of passive techniques compared with active techniques that require extra devices.

Such passive techniques include stereo vision, trinocular vision (Lehel et al. 1999, Kim 2004, Farag 2004), and many monocular shape-from-X techniques, e.g. 3D shape from texture, motion parallax, focus, defocus, shadows, shading, specularities, occluding contours, and other surface discontinuities. The problem is that recovering 3D information from a single 2D image is an ill-posed problem (Papadopoulos 2001). Stereo vision is still the single passive cue that gives reasonable accuracy. Human has two eyes, and precisely because of the way the world is projected differently onto the eyes, human is able to obtain the relative distances of objects. The setup of a stereo machine vision system also has two cameras, separated by a baseline distance b. The 3D world point may be measured by the two projection equations, in a way that is analogous to the way the human eyes work. To interpret disparity between images, the matching problem must be solved, which has been formulated as an ill-posed problem in a general context and which anyway is a task difficult to automate. This correspondence problem results in an inaccurate and slow process and reduces its usefulness in many practical applications (Blais 2004). The other major drawback of this passive approach is that it requires two cameras and it cannot be used on un-textured surfaces which are common for industrially manufactured objects. The requirement of ambient light conditions is also critical in passive visual sensing. The advantage of stereo vision is that it is very convenient to implement and especially suitable for natural environments. A few applications are illustrated in Figs. 2.1 to 2.3.

The structure-from-motion algorithms solve the following problem: given a set of tracked 2D image features captured by a moving camera, find the 3D positions and orientations of the corresponding 3D features (structure) as well as the camera motion. Pose estimation, on the other hand, solves the problem of finding the position and orientation of a camera given correspondences between 3D and 2D features. In both problems two-dimensional line features are advantageous because they can be reliably extracted and are prominent in man-made scenes. Taylor and Kriegman (1995) minimized a nonlinear objective function with respect to camera rotation, camera translation and 3D lines parameters. The objective function measures the deviation of the projection of the 3D lines on the image planes from the extracted image lines. This method provides a robust solution to the high-dimensional non-linear estimation problem. Fitzgibbon and Zisserman (1998) also worked towards the automatic construction of graphical models of scenes when the input was a sequence of closely spaced images. The point features were matched in triples of consecutive images and the fundamental matrices were estimated from pairs of images. The projective reconstruction and camera pose estimation was upgraded to a Euclidean one by means of auto-calibration techniques (Pollefeys et al. 1998). Finally, the registration of image coordinate frames was based on the algorithm of iterative closest points (Besl and Mckay 1992).

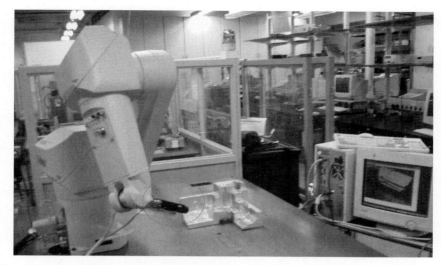

Fig. 2.1. Stereo vision for industrial robots

Fig. 2.2. Mars Rover in 3D (NASA mission in 2003–2004) (Pedersen 2003, Miller 2003, Madison 2006, Deen and Lore 2005)

Fig. 2.3. MR-2 (Prototype of Chinese Moon Explorer in 2007–2008)

2.1.2 Active Visual Sensing

In contrast to passive visual sensing, the other class of visual sensing techniques is called active visual sensing. For the above cases of passive techniques (that use ambient light), only visible features with discernable texture gradients like on intensity edges are measured. For the example of the stereo setup, there is a corresponding problem. Matching corresponding points is easy if the difference in position and orientation of the stereo views is small, whereas it is difficult if the difference is large. However, the accuracy of the 3D reconstruction tends to be poor when the difference in position and orientation of the stereo views is small. To overcome the shortcomings of passive sensing, active sensing techniques have been developed in the recent years. These active systems usually do not have the correspondence problem and can measure with a very high precision.

By active sensing, an external projecting device (e.g. laser or LCD/DLP projector) is used to actively emit light patterns that are reflected by the scene and detected by a camera. That is to say they rely on probing the scene in some way rather than relying on natural lighting. Compared with the passive approach, active visual sensing techniques are in general more accurate and reliable.

Generally active 3D vision sensors can resolve most of the ambiguities and directly provide the geometry of an object or an environment. They require minimal operator assistance to generate the 3D coordinates. However, with laser-based approaches, the 3D information becomes relatively insensitive to background illumination and surface texture. Therefore, active visual sensing is ideal for scenes that do not contain sufficient features. Since it requires lighting control, it is usually suitable for indoor environments and both camera and projector need to be pre-calibrated.

Typically, properly formatted light, or another form of energy, is emitted in the direction of an object, reflected on its surface and received by the sensor; the distance to the surface is calculated using triangulation or time-of-flight (Papadopoulos 2001). Typical triangulation-based methods include single/multi-point projection, line projection, fringe and coded pattern projection, and moire effect (Figs. 2.4–2.6). Typical time-of-flight based methods are interferometers and laser range finders.

Moire devices work on the principle that: effectively projecting a set of fringe patterns on a surface using an interference technique, tracking the contours of the fringes allows the range to be deduced. Systems that use point projection, line scanning, and moiré effect are highly accurate, but can be slow. Moire devices are best suited to digitizing surfaces with few discontinuities.

Interferometers work on the principle that: if a light beam is divided into two parts (reference and measuring) that travel different paths, when the beams are combined together interference fringes are produced. With such devices, very small displacements can be detected. Longer distances can also be measured with low measurement uncertainty (by counting wavelengths).

For laser range finders, the distance is measured as a direct consequence of the propagation delay of an electromagnetic wave. This method usually provides good distance precision with the possibility of increasing accuracy by means of longer measurement integration times. The integration time is related to the number of samples in each measurement. The final measurement is normally an average of the sample measures, decreasing therefore the noise associated to each single measure. Spatial resolution is guaranteed by the small aperture and low divergence of the laser beam (Sequeira et al. 1995, 1996, 1999). Basically laser range finders work in two different techniques: pulsed wave and continuous wave. Pulsed wave techniques are based on the emission and detection of a pulsed laser beam. A short laser pulse is emitted at a given frequency and the time elapsed between the emission and the received echo is measured. This time is proportional to the distance from the sensor to the nearest object. In a continuous wave laser ranging system, rather than using a short pulse, a continuous laser beam modulated with a reference waveform is emitted and the range is determined as a result of the comparison of the emitted and received laser beams. This type of system can use either amplitude modulation (e.g. sinusoidal signal) or frequency modulation.

Among various 3D range data acquisition techniques in computer vision, the structured light system with coded patterns is based on active triangulation. A very simple technique to achieve depth information with the help of structured light is to scan a scene with a laser plane and to detect the location of the reflected stripe. The depth information can be computed out of the distortion along the detected profile. More complex techniques of structured light project multiple stripes (Fig. 2.7) or a pattern of grids at once onto the scene. In order to distinguish between stripes or grids they are coded either with different brightness or different colors (Fig. 2.8) (e.g. Coded Light Approach (Inokuchi et al. 1984, Stahs and Wahl 1992) and unique color encoding method). The structured light systems, as well as laser range finders, map directly the acquired data into a 3D volumetric model having thus the ability to avoid the correspondence problem associated with passive sensing techniques. Indeed, scenes with no textural details can be easily modeled. A drawback with the technique of coded stripes is that because each projection direction is associated with a code word, the measurement resolution is low. Fortunately, when this approach is combined with a phase-shift approach, a theoretically infinite height resolution can be obtained. For available products, Fig. 2.9 illustrates some examples of 3D laser scanners and Fig. 2.10 illustrates some examples of 3D Structured Light System.

Therefore, although there are many types of vision sensors available to measure object models by either passive or active methods, structured-light is one of the most important methods due to its many advantages compared with other methods, and thus it is successfully used in many areas for recovering 3D information of an industrial object. This chapter considers typical setups of the structured light system for active visual sensing, using stripe light vision or color-encoded vision. Their system configurations and measurement principles are presented in the following sections.

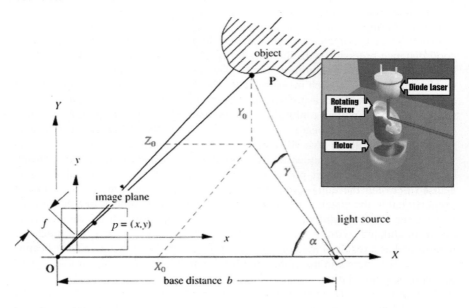

Fig. 2.4. Light spot projection

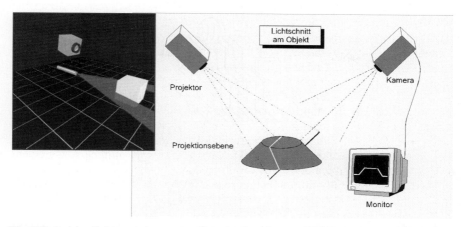

Fig. 2.5. A stripe light scanning system (Intersecting the projection ray with an additional ray or plane will lead to a unique reconstruction of the object point.)

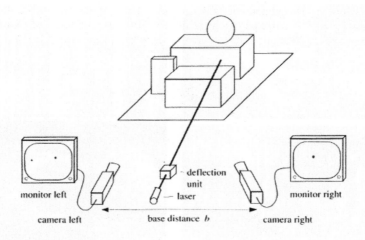

Fig. 2.6. Single spot stereo analysis

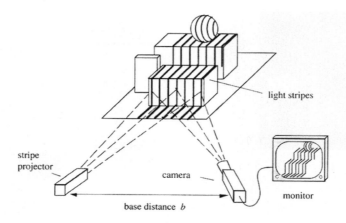

Fig. 2.7. Stripe light vision system

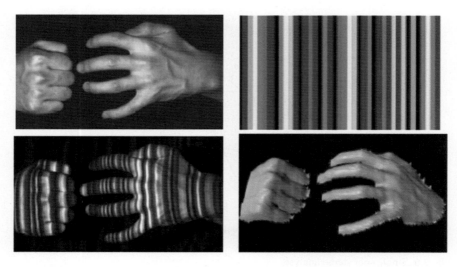

Fig. 2.8. Coded structured light vision: project a light pattern into a scene and analyze the modulated image from the camera

Fig. 2.9. Examples of 3D laser scanners

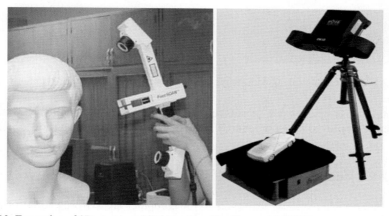

Fig. 2.10. Examples of 3D structured light system (FastScan and OKIO)

2.2 3D Sensing by Stereo Vision Sensors

2.2.1 Setup with Two Cameras

Binocular stereo vision is an important way of getting depth (3D) information about a scene from two 2-D views of the scene. Inspired by the vision mechanism of humans and animals, computational stereo vision has been extensively studied in the past 30 years, for measuring ranges by triangulation to selected locations imaged by two cameras. However, some difficulties still exist and have to be researched further. The figure illustrated below contains several examples of mobile

Fig. 2.11. The mobile robots with stereo vision setup at the University of Hamburg

robots that use stereo vision sensors for understanding the 3D environment, which are currently employed in our laboratory at the University of Hamburg (Fig. 2.11).

2.2.2 Projection Geometry

In a stereo vision system, the inputs to the computer are 2D-projections of the 3D object. The vision task is to reconstruct 3D world coordinates according to such 2D projected images, so we must know the relationship between the 3D objective world and 2D images (Fig. 2.12), namely the projection matrix. A camera is usually described using the pinhole model and the task of calibration is to confirm the projection matrix. As we know, there exists a collineation which maps the projective space to the camera's retinal plane: $P^3 \rightarrow P^2$. Then the coordinates of a 3D point $\mathbf{X} = [X, Y, Z]^T$ in a Euclidean world coordinate system and the retinal image coordinates $\mathbf{x} = [u, v]^T$ are related by the following (2.1).

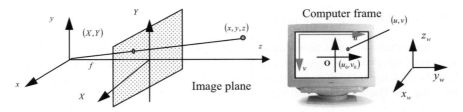

Fig. 2.12. The projection geometry: from 3D world to 2D image

$$\lambda \begin{bmatrix} u \\ v \\ 1 \end{bmatrix} = \begin{bmatrix} f_x & s & u_0 \\ & f_y & v_0 \\ & & 1 \end{bmatrix} \begin{bmatrix} 1 & & \\ & 1 & \\ & & 1 \end{bmatrix} \begin{bmatrix} \mathbf{R} & \mathbf{t} \\ 0_3^\mathsf{T} & 1 \end{bmatrix} \begin{bmatrix} X \\ Y \\ Z \\ 1 \end{bmatrix}, \tag{2.1}$$

where λ is a scale factor, $c = [u_0, v_0]^\mathsf{T}$ is the principal point, f_x and f_y are focal lengths, s is the skew angle, and $\mathbf{R}$ and $\mathbf{t}$ are external or extrinsic parameters. $\mathbf{R}$ is the 3×3 rotation matrix which gives axes of the camera in the reference coordinate system and $\mathbf{t}$ the translation in the X, Y and Z directions representing the camera center in the reference coordinate system (Henrichsen 2000). Of course, it is the same reference coordinate system in both views of the stereo couple.

$$\mathbf{R} = \mathbf{R_x}(\theta_x)\mathbf{R}_y(\theta_y)\mathbf{R}_z(\theta_z) = \begin{bmatrix} r_{11} & r_{12} & r_{13} \\ r_{21} & r_{22} & r_{23} \\ r_{31} & r_{32} & r_{33} \end{bmatrix}, \tag{2.2}$$

$$\mathbf{t} = [T_x \ T_y \ T_z]^\mathsf{T}. \tag{2.3}$$

Equation (2.1) can be expressed as

$$\lambda \mathbf{x} = \mathbf{M} \mathbf{X}, \tag{2.4}$$

where $\mathbf{x} = [u, v, 1]^\mathsf{T}$ and $\mathbf{X} = [X, Y, Z, 1]^\mathsf{T}$ are the homogeneous coordinates of spatial vectors, and $\mathbf{M}$ is a 3×4 matrix, called the perspective projection, representing the collineation: $P^3 \to P^2$ (Henrichsen 2000).

The first part of the projection matrix in the collineation (2.1), denoted by $\mathbf{K}$, contains the intrinsic parameters of the camera used in the imaging process. This matrix is used to convert between the retinal plane and the actual image plane. In a normal camera, the focal length mentioned above does not usually correspond to 1. It is also possible that the focal length changes during an entire imaging process, so that for each image the camera calibration matrix needs to be reestablished (denoted as *recalibration* in a following section).

2.2.3 3D Measurement Principle

If the value of the same point in computer image coordinate shoot by two cameras can be obtained, the world coordinates of the points can be calculated through the projection of two cameras (Fig. 2.13). Then four equations can be obtained from the two matrix formulas and the world coordinates of the point can be calculated (Ma and Zhang, 1998).

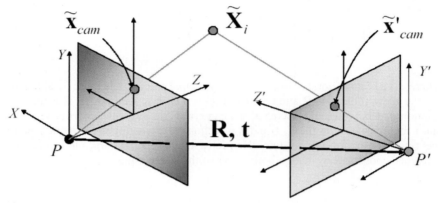

Fig. 2.13. 3D measurement by stereo vision sensors

For the two cameras, we have

$$\lambda_1 \mathbf{x}_1 = \mathbf{M}_1 \mathbf{X} \text{ and } \lambda_2 \mathbf{x}_2 = \mathbf{M}_2 \mathbf{X}, \tag{2.5}$$

or

$$Z_{c1} \begin{bmatrix} u_1 \\ v_1 \\ 1 \end{bmatrix} = \begin{bmatrix} m_{11}^1 & m_{12}^1 & m_{13}^1 & m_{14}^1 \\ m_{21}^1 & m_{22}^1 & m_{23}^1 & m_{24}^1 \\ m_{31}^1 & m_{32}^1 & m_{33}^1 & m_{34}^1 \end{bmatrix} \begin{bmatrix} X \\ Y \\ Z \\ 1 \end{bmatrix}$$

$$Z_{c2} \begin{bmatrix} u_2 \\ v_2 \\ 1 \end{bmatrix} = \begin{bmatrix} m_{11}^2 & m_{12}^2 & m_{13}^2 & m_{14}^2 \\ m_{21}^2 & m_{22}^2 & m_{23}^2 & m_{24}^2 \\ m_{31}^2 & m_{32}^2 & m_{33}^2 & m_{34}^2 \end{bmatrix} \begin{bmatrix} X \\ Y \\ Z \\ 1 \end{bmatrix} \tag{2.6}$$

The two uncertain numbers can be removed and (2.6) becomes

$$\begin{aligned}
(u_1 m_{31}^1 - m_{11}^1)X + (u_1 m_{32}^1 - m_{12}^1)Y + (u_1 m_{33}^1 - m_{13}^1)Z &= m_{14}^1 - u_1 m_{34}^1 \\
(v_1 m_{31}^1 - m_{21}^1)X + (v_1 m_{32}^1 - m_{22}^1)Y + (v_1 m_{33}^1 - m_{23}^1)Z &= m_{24}^1 - v_1 m_{34}^1 \\
(u_2 m_{31}^2 - m_{11}^2)X + (u_2 m_{32}^2 - m_{12}^2)Y + (u_2 m_{33}^2 - m_{13}^2)Z &= m_{14}^2 - u_2 m_{34}^2 \\
(v_2 m_{31}^2 - m_{21}^2)X + (v_2 m_{32}^2 - m_{22}^2)Y + (v_2 m_{33}^2 - m_{23}^2)Z &= m_{24}^2 - v_2 m_{34}^2
\end{aligned} \tag{2.7}$$

or

$$\begin{bmatrix} q_{11} & q_{12} & q_{13} \\ q_{21} & q_{22} & q_{23} \\ q_{31} & q_{32} & q_{33} \\ q_{41} & q_{42} & q_{43} \end{bmatrix} \begin{bmatrix} X \\ Y \\ Z \end{bmatrix} = \begin{bmatrix} b_1 \\ b_2 \\ b_3 \\ b_4 \end{bmatrix}, \tag{2.8}$$

$$\mathbf{Q}[X\ Y\ Z]^T = \mathbf{B}$$

In this linear system (2.8), the only three unknowns (X, Y, Z) can be solved simply by:

$$[X \; Y \; Z]^T = (\mathbf{Q}^T\mathbf{Q})^{-1}\mathbf{Q}^T\mathbf{B} \qquad (2.9)$$

In practice, correction of lens distortion and epipolar geometry for feature matching should be considered for improving the efficiency and accuracy of 3D reconstruction. The epipolar plane contains the three-dimensional point of interest, the two optical centers of the cameras, and the image points of the point of interest in the left and right images. An epipolar line is defined by the intersection of the epipolar plane with image planes of the left and right cameras. The epipole of the image is the point where all the epipolar lines intersect. More intensive technology for stereo measurement is out of the scope of this book, but can be found in many published contributions.

2.3 3D Sensing by Stripe Light Vision Sensors

Among the active techniques, the structured-light system features high quality and reliability for 3D measurement. It may be regarded as a modification of static binocular stereo. One of the cameras is replaced by the projector which projects (instead of receives) onto the scene a sheet of light (or multiple sheets of light simultaneously). The simple idea is that once the perspective projection matrix of the camera and the equations of the planes containing the sheets of light relative to a global coordinate frame are computed from calibration, the triangulation for computing the 3D coordinates of object points simply involves finding the intersection of a ray (from the camera) and a plane (from the light pattern of the projector). A controllable LCD (Liquid Crystal Display) or DLP (Digital Light Processing) projector is often used to illuminate the surface with particular patterns. It makes it possible for all the surfaces in the camera's field of view to be digitized in one frame, and so is suitable for measuring objects at a high field rate.

2.3.1 Setup with a Switchable Line Projector

The active visual sensor considered in this section consists of a projector, which is a switchable LCD line projector in this study, to cast a pattern of light stripes onto the object and a camera to sense the illuminated area as shown in Fig. 2.14. The 3D measurement is based on the principle of triangulation. If a beam of light is cast, and viewed obliquely, the distortions in the beam line can be translated into height variations. The correspondence problem is avoided since the triangulation is carried out by intersecting the two light rays generated from the projector and seen by the camera.

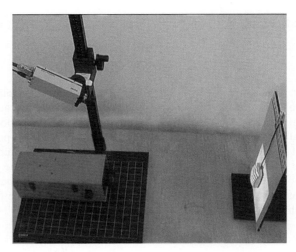

Fig. 2.14. Setup of stripe light vision system

The projector is controllable by a computer to select a specific light pattern. All the patterns are pre-designed with light and dark stripes and switchable during the operation.

2.3.2 Coding Method

In structured light systems, the light coding method is used as a technique to solve the correspondence problem (Batlle et al. 1998, Salvi et al. 2004). 3D measurement by structured lighting is based on the expectation of precise detection of the projected light patterns in the acquired images. The 3D coordinates can be triangulated directly as soon as the sensor geometry has been calibrated and the light pattern is located in the image.

For such systems as shown in Fig. 2.14, a Coded Light Approach is most suitable for space-encoding and position detection when using a switchable LCD or mask. It is also an alternative approach for avoiding the scanning of the light and it requires only a small number of images to obtain a full depth-image. This can be achieved with a sequence of projections using a set of switchable lines (light or dark) on the LCD device. All the lines are numbered from left to right.

In a so-called gray-coding (Inokuchi et al. 1984, Stahs and Klahl 1992), adjacent lines differ by exactly one bit leading to good fault tolerance. Using a projector, all lines (e.g. 512 switchable lines) may be encoded with several bits. This can be encoded in 10 projected line images. One bit of all lines is projected at a time. A bright line represents a binary '0', a dark line a '1'. All object points illuminated by the same switchable line see the same sequence of bright and dark illuminations. After a series of exposures, the bit-plane stack contains the encoded number of the corresponding lines in the projector. This is the angle in encoded format. The angle α is obtained from the column address of each pixel. Thus all the

Fig. 2.15. An example of the coding method

information needed to do the triangulation for each pixel is provided by the x-address and the contents of the bit-plane stack. Using look-up-tables can generate a full 3D image within a few seconds. With such a setup, the depth resolution can be further increased using the phase-shift method or the color-coded method.

Figure 2.15 illustrates an example of the coding method. The lines are numbered from left to right. They are called Gray-Code, although they are in binary patterns. Using a controllable projector with 2^n switchable lines, all lines may be encoded with $n+1$ bits, and projected with $n+1$ images.

2.3.3 Measurement Principle

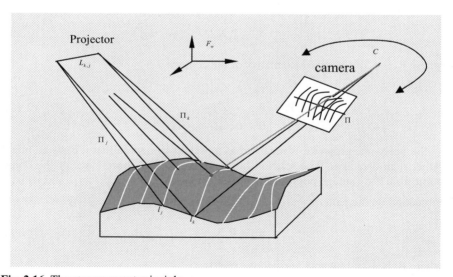

Fig. 2.16. The measurement principle

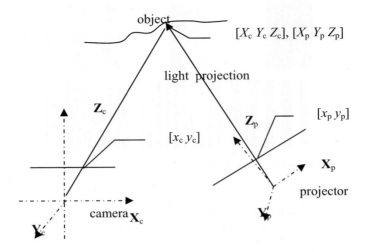

Fig. 2.17. The coordinates in the system

Figure 2.17 illustrates the measurement principle in the stripe light system and Fig. 2.18 illustrates the representation of point coordinates. For the camera, the relationship between the 3D coordinates of an object point from the view of the camera $\mathbf{X}_c = [X_c \quad Y_c \quad Z_c \quad 1]^T$ and its projection on the image $\mathbf{x}_c = [\lambda x_c \quad \lambda y_c \quad \lambda]^T$ is given by

$$\mathbf{x}_c = \mathbf{P}_c \mathbf{X}_c, \tag{2.10}$$

where $\mathbf{P}_c$ is a 3×4 perspective matrix of the camera:

$$\mathbf{P}_c = \begin{bmatrix} v_x & k & x_{c0} & 0 \\ 0 & v_y & y_{c0} & 0 \\ 0 & 0 & 1 & 0 \end{bmatrix} \tag{2.11}$$

Similarly, the projector is regarded as a pseudo-camera in that it casts an image rather than detects it. The relationship between the 3D coordinates of the object point from the vantage point of the projector $\mathbf{X}_p = [X_p \quad Y_p \quad Z_p \quad 1]^T$ and its back projection on the pattern sensor (LCD/DMD) $\mathbf{x}_p = [\kappa x_p \quad \kappa]^T$ is

$$\mathbf{x}_p = \mathbf{P}_p \mathbf{X}_p \tag{2.12}$$

where $\mathbf{P}_p$ is a 2×4 inverse perspective matrix:

$$\mathbf{P}_p = \begin{bmatrix} v_p & 0 & x_p^0 & 0 \\ 0 & 0 & 1 & 0 \end{bmatrix} \tag{2.13}$$

The relationship between the camera view and the projector view is given by

$$\mathbf{X}_p = \mathbf{M}\mathbf{X}_c, \quad \mathbf{M} = \mathbf{R}_\theta \mathbf{R}_\alpha \mathbf{R}_\beta \mathbf{T},$$ (2.14)

in which $\mathbf{R}_\theta$, $\mathbf{R}_\alpha$, $\mathbf{R}_\beta$, and $\mathbf{T}$ are 4×4 matrices standing for 3-axis rotation and translation.

Substituting (2.14) into (2.12) yields

$$\mathbf{x}_p = \mathbf{P}_p \mathbf{M}\mathbf{X}_c.$$ (2.15)

Let $$\mathbf{H} = \mathbf{P}_p \mathbf{M} = \begin{bmatrix} \mathbf{r}_1 \\ \mathbf{r}_2 \end{bmatrix},$$ (2.16)

where $\mathbf{r}_1$ and $\mathbf{r}_2$ are 4-dimensional row vectors. Equation (2.15) becomes

$$(x_p \mathbf{r}_2 - \mathbf{r}_1)\mathbf{X}_c = 0.$$ (2.17)

Combining (2.10) and (2.17) gives

$$\begin{bmatrix} \mathbf{P}_c \\ x_p \mathbf{r}_2 - \mathbf{r}_1 \end{bmatrix} \mathbf{X}_c = \begin{bmatrix} \mathbf{x}_c \\ 0 \end{bmatrix}$$ (2.18)

or

$$\mathbf{Q}\mathbf{X}_c = \mathbf{x}_{c+}$$ (2.19)

where $\mathbf{Q}$ is a 4 by 4 matrix.

Then the three-dimensional world position of a point on the object surface can be determined by

$$\mathbf{X}_c = \mathbf{Q}^{-1}\mathbf{x}_{c+}$$ (2.20)

From the above equations, the 3D object can be uniquely reconstructed if we know the matrix $\mathbf{Q}$ that contains 13 parameters from the two perspective matrices $\mathbf{P}_c$ and $\mathbf{P}_p$ and the coordinate transformation matrix $\mathbf{M}$. This means, once the perspective matrices of the camera and projector relative to a global coordinate frame are given from calibration, the triangulation for computing the 3D coordinates of object points simply involves finding the intersection of a ray from the camera and a stripe plane from the projector.

2.4 3D Sensor Reconfiguration and Recalibration

Since the objects may be different in sizes and distances and the task requirements may also be different for different applications, a structure-fixed vision sensor does not work well in such cases. A reconfigurable sensor, on the other hand, can change its structural parameters to adapt itself to the scene to obtain maximum 3D information from the environment. If reconfiguration occurs, the sensor should be capable of self-recalibration so that 3D measurement can follow immediately.

2.4.1 The Motivation for Sensor Reconfiguration and Recalibration

In an active visual system, since the sensor needs to move from one place to another for performing a multi-view vision task, a traditional vision sensor with fixed structure is often inadequate for the robot to perceive the object features in an uncertain environment as the object distance and size are unknown before the robot sees it. A dynamically reconfigurable sensor can help the robot to control the configuration and gaze at the object surfaces. For example, with a structured light system, the camera needs to see the object surface illuminated by the projector, to perform the 3D measurement and reconstruction task. Active calibration means that the vision sensor is reconfigurable during runtime to fit in the environment and can perform self-recalibration in need before 3D perception.

The concept of self-calibration in stereo vision and camera motion has been studied for more than ten years and there are many useful outputs. It is an attempt to overcome the problem of manual labor. For example, using the invariant properties of calibration matrix to motions, Dias et al. (1991) proposed an optimization procedure for recalibration of a stereo vision sensor mounted on a robot arm. The technique for self-recalibration of varying internal and external parameters of a camera was explored in (Zomet et al. 2001). The issues in dynamic camera calibration were addressed to deal with unknown motions of the cameras and changes in focus (Huang and Mitchell 1995). A method for automatic calibration of cameras was explored by tracking a set of world points (Wei et al. 1998). Such self-calibration techniques normally require a sequence of images to be captured via moving the camera or the target (Kang 2000). With some special setups, two views can also be sufficient for such a calibration (Seo and Hong 2001). All these are passive methods to calibrate the sensor with some varying intrinsic parameters.

For structured light vision systems, most existing methods are still based on static and manual calibration. That is, during the calibration and 3D reconstruction, the vision sensor is fixed in one place. The calibration target usually needs to be placed at several accurately known or measured positions in front of the sensor (DePiero and Trivedi 1996, Huynh 1997, Sansoni et al. 2000). With these traditional methods, the system must be calibrated again if the vision sensor is moved or the relative pose between the camera and the projector is changed. For the active vision system working in an unknown environment, changes of the position and configuration of the vision sensor become necessary. Frequent recalibrations in using such a system are tedious tasks. The recalibration means that the sensor has been calibrated before installation on the robot, but it needs to be calibrated again as its relative configuration is changing.

However, only a few related works can be found on self-calibration of a structured-light system. Furthermore, the self-calibration methods for a passive camera cannot be directly applied to an active vision system which includes an illumination system using structured light in addition to the traditional vision sensor. Among the previous self-calibration works on structured light systems, a self-reference method (Hébert 2001) was proposed by Hebert to avoid using the external calibrating device and manual operations. A set of points was projected on the scene and was detected by the camera to be used as reference in the calibration of a hand-held range sensor. With a cubic frame, Chu et al. proposed a calibration-free approach for recovering unified world coordinates (Chu et al. 2001). Fofi et al. discussed the problem in self-calibrating a structured light sensor (Fofi et al. 2001). A stratified reconstruction method based on Euclidean constraints by projection of a special light pattern was given. However, the work was based on the assumption that "projecting a square onto a planar surface, the more generic quadrilateral formed onto the surface is a parallelogram". This assumption is questionable. Consider an inclined plane placed in front of the camera or projector. Projecting a square on it forms an irregular quadrangle instead of a parallelogram as the two line segments will have different lengths on the image plane due to their different distances to the sensor. Jokinen's method (Jokinen 1999) of self-calibration of light stripe systems is based on multiple views. The object needs to be moved by steps and several maps are acquired for the calibration. The registration and calibration parameters are obtained by matching the 3D maps via least errors. The limitation of this method is that it requires a special device to hold and move the object.

This chapter studies the problems of "self-calibration" and "self-recalibration" of active sensing systems. Here self-recalibration deals with situations where the system has been initially calibrated but needs to be automatically calibrated again due to a changed relative pose. Self-calibration refers to cases where the system has never been calibrated and none of the sensor's parameters including the focal length and relative pose are known. Both of them do not require manual placements of a calibration target. Although the methods described later in this book are mainly concerned with the former situation, they can also be applied to the latter case if the focal lengths of the camera and the projector can be digitally controlled by a computer.

The remainder part of this section investigates the self-recalibration of a dynamically reconfigurable structured light system. The intrinsic parameters of the projector and camera are considered as constants, but the extrinsic transformation between light projector and camera can be changed. A distinct advantage of the method is that neither an accurately designed calibration device nor the prior knowledge of the motion of the camera or the scene is required during the recalibration. It only needs to capture a single view of the scene.

2.4.2 Setup of a Reconfigurable System

For the stripe light vision system (Fig. 2.16), to make it adaptable to different objects/scenes to be sensed, we incorporated two degrees of freedom of relative motion in the system design, i.e. the orientation of the projector (or the camera) and its horizontal displacement. This 2DOF reconfiguration is usually adequate for many practical applications. Where more DOFs are necessary, the recalibration issues will be addressed in the next section. In Fig. 2.18, the camera is fixed whereas the projector can be moved on a horizontal track and rotated around the y-axis. The x-z plane is orthogonal to the plane of the projected laser sheet. The 3D

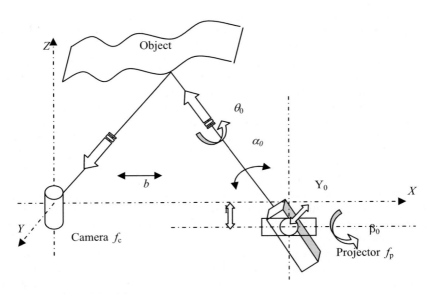

Fig. 2.18. A reconfigurable system

coordinate system is chosen based on the camera (or the projector) center and its optic axis.

For such a reconfigurable system, the two perspective matrices, which contain the intrinsic parameters such as the focal lengths (v_c, v_p) and the optical centers (x_{c0}, y_{c0}, x_{p0}), can be determined in the initial calibration stage. Based on the formulation in the previous section, the dynamic recalibration task is to determine the relative matrix **M** in (2.14) between the camera and the projector. There are six parameters, i.e.

$$\mathbf{u} = [\theta \quad \alpha \quad \beta \quad X_0 \quad Y_0 \quad Z_0] \tag{2.21}$$

Since the system considered here has two DOFs, only two of the six parameters are variable while the other four are constants which can be known from the initial calibration. If the X-Z plane is not perpendicular to the plane of the projected laser sheet, its angle Φ can also be identified at this stage. As the angle $\theta_0 = (90° - \Phi)$ is small and the image can be rectified by rotating the corresponding angle accordingly during recalibration, it can be assumed that $\theta_0 = 0$. The displacement in the y-direction between the camera center and the projector center, Y_0, and the rotation angle β_0 are also small in practice. They do not affect the 3D reconstruction as the projected illumination consists of vertical line stripes. Therefore, we may assume that $Y_0 = 0$ and $\beta_0 = 0$. Thus, the unknown parameters are reduced to only two (α_0 and b) for the dynamic recalibration. Here h is a constant and α_0 and b have variable values depending on the system configuration.

For such a 2DOF system, the triangulation (2.18) for determining the 3D position of a point on the object surface is then simplified as (see Fig. 2.18)

$$[X_c \ Y_c \ Z_c] = \frac{b - h\cot(\alpha)}{v_c \cot(\alpha) + x_c} \ [x_c \ y_c \ v_c],$$ (2.22)

where v_c is the distance between the camera sensor and the optical center of the lens, $\alpha = \alpha(i) = \alpha_0 + \alpha_p(i)$ is the projection angle, and

$$\alpha_p(i) = \tan^{-1}(\frac{x_p(i)}{v_p}),$$ (2.23)

where i is the stripe index and $x_p(i)$ is the stripe coordinate on the projection plane $x_p(i) = i \times$ stripe width $+ x_p(0)$.

If the projector's rotation center is not at its optical center, h and b shall be replaced by:

$$h' = h - r_0 \sin(\alpha_0) \ \text{and} \ b' = b - r_0 \cos(\alpha_0)$$ (2.24)

where r_0 is the distance between the rotational center and the optical center (Fig. 2.19). h and r_0 can be determined during the initial static calibration. Figure 2.20 illustrates the experimental device which is used to determine the camera's rotation axis.

For the reconfigurable system, the following sections show a self-recalibration method which is performed if and when the relative pose between the camera and the projector is changed. The unknown parameters are determined automatically by using an intrinsic cue, the geometrical cue. It describes the intrinsic relationship between the stripe locations on the camera and the projector. It forms a geometrical constraint and can be used to recognize the unknown parameters of the vision system.

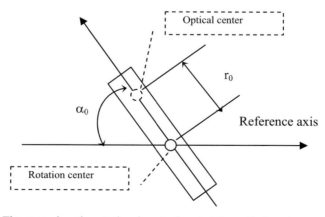

Fig. 2.19. The case when the rotational center is not at the optical center

Fig. 2.20. The device to calibrate the rotational center

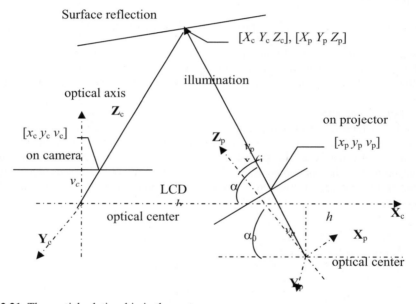

Fig. 2.21. The spatial relationship in the system

2.4.3 Geometrical Constraint

Assume a straight line in the scene which is expressed in the camera coordinate system and projected on the X-Z plane:

$$Z_c = C_1 X_c + C_2. \tag{2.25}$$

The geometrical constraint between projecting and imaging of the scene line is obtained by substituting (2.22) into (2.25):

$$[b - h\cot(\alpha)] \, (v_c - C_1 x_c) - C_2 \, [v_c\cot(\alpha) + x_c] = 0. \tag{2.26}$$

If $h = 0$, the above can be simplified as

$$b(v_c - C_1 x_c) - C_2[v_c\cot(\alpha_0 + \alpha_{pi}) + x_c] = 0. \tag{2.27}$$

For two point pairs (x_{ci}, α_{pi}) and (x_{cj}, α_{pj}),

$$C_1[x_{cj} \frac{v_c \cot(\alpha_0 + \alpha_{pi}) + x_{ci}}{v_c \cot(\alpha_0 + \alpha_{pj}) + x_{cj}} - x_{ci}] = v_c[\frac{v_c \cot(\alpha_0 + \alpha_{pi}) + x_{ci}}{v_c \cot(\alpha_0 + \alpha_{pj}) + x_{cj}} - 1]. \tag{2.28}$$

Denote $F_{ij} = \dfrac{v_c \cot(\alpha_0 + \alpha_{pi}) + x_{ci}}{v_c \cot(\alpha_0 + \alpha_{pj}) + x_{cj}}$. The coordinates of four points yield

$$\frac{x_{cj}F_{ij} - x_{ci}}{x_{cl}F_{kl} - x_{ck}} = \frac{F_{ij} - 1}{F_{kl} - 1}. \tag{2.29}$$

With (2.29), if the locations of four stripes are known, the projector's orientation α_0 can be determined when assuming $h = 0$.

If $h \neq 0$, from (2.27), the parameters v_c, h, and v_p are constants that have been determined at the initial calibration stage. $x_c = x_{ci} = x_c(i)$ and $\alpha_{pi} = \alpha_p(i)$ are known coordinates on the sensors. Therefore, α_0, b, C_1, and C_2 are the only four unknown constants and their relationship can be defined by three points.

Denote $A_0 = \tan(\alpha_0)$ and $A_i = \tan(\alpha_{pi})$. The projection angle of an illumination stripe is (illustrated in Fig. 2.22)

$$\cot(\alpha_0 + \alpha_{pi}) = \frac{1 - \tan(\alpha_0)\tan(\alpha_{pi})}{\tan(\alpha_0) + \tan(\alpha_{pi})} = \frac{1 - A_0 A_i}{A_0 + A_i} = \frac{v_p - A_0 x_p}{v_p A_0 + x_p}, \tag{2.30}$$

where $x_p = x_p(i)$ is the stripe location on the projector's LCD. The x-coordinate value of the i th stripe, $x_p(i)$, can be determined by the light coding method. The stripe coordinate x_p and the projection angle α_{pi} are illustrated in Figs. 2.21 and 2.22.

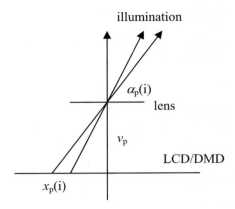

Fig. 2.22. The projection angle

Equation (2.27) may be written as

$$
\begin{aligned}
(bA_0 - C_2 - h)v_c v_p + (hC_1 - bC_1 A_0 - C_2 A_0)v_p x_c + \\
(b + A_0 C_2 + hA_0)v_c x_p - (C_2 + bC_1 + hC_1 A_0)x_c x_p = 0
\end{aligned}
\tag{2.31}
$$

or

$$
x_c = W_1 + W_2 x_p + W_3 (x_c x_p),
\tag{2.32}
$$

where

$$
W_1 = \frac{V_2 V_3 - V_1 V_4}{V_2 V_4} - \frac{V_3}{V_4}, \; W_2 = -\frac{V_3}{V_2}, \; \text{and} \; W_3 = \frac{V_2}{V_4}
\tag{2.33}
$$

$$
V_1 = bA_0 - C_2 - h,
\tag{2.34}
$$

$$
V_2 = b + A_0 C_2 + hA_0,
\tag{2.35}
$$

$$
V_3 = hC_1 - bC_1 A_0 - C_2 A_0,
\tag{2.36}
$$

$$
V_4 = C_2 + bC_1 + hC_1 A_0,
\tag{2.37}
$$

Equation (2.32) is the relationship between the stripe locations on the camera and the projector and is termed the geometrical constraint.

2.4.4 Rectification of Stripe Locations

Within a view of the camera, there can be tens or hundreds of stripes from the scene. The stripes' coordinates (x_c, x_p) on the image and the projector should satisfy (2.32) in theory. In practice, however, the coordinates (x_c) obtained from the image processing may not satisfy this constraint, due to the existence of noise. To reduce the effect of noise and improve the calibration accuracy, the stripe locations on the image can be rectified by using a curve fitting method.

Let the projection error be

$$Q_{err}(W_1, W_2, W_3) = \sum_{i=1}^{m}[x_c'(i) - x_c(i)]^2 = \sum_{i=1}^{m}[W_1 + W_2 x_p + W_3(x_c x_p) - x_c]^2 \cdot \quad (2.38)$$

Then W_1, W_2, and W_3 may be obtained by minimizing the projection error Q_{err} with respect to W_k

$$\frac{\partial Q_{err}}{\partial W_k} = 0, \quad k = 1, 2, 3. \quad (2.39)$$

Using (2.38) in (2.39) gives

$$\begin{bmatrix} m & \sum_{i=1}^{m} x_p(i) & \sum_{i=1}^{m} x_c(i)x_p(i) \\ \sum_{i=1}^{m} x_p(i) & \sum_{i=1}^{m} x_p^2(i) & \sum_{i=1}^{m} x_c(i)x_p^2(i) \\ \sum_{i=1}^{m} x_c(i)x_p(i) & \sum_{i=1}^{m} x_c(i)x_p^2(i) & \sum_{i=1}^{m} x_c^2(i)x_p^2(i) \end{bmatrix}\begin{bmatrix} W_1 \\ W_2 \\ W_3 \end{bmatrix} =$$

$$\left[\sum_{i=1}^{m} x_c(i) \quad \sum_{i=1}^{m} x_p(i)x_c(i) \quad \sum_{i=1}^{m} x_p(i)x_c^2(i)\right]^{\mathrm{T}} \quad (2.40)$$

$$\text{Or } \mathbf{GW} = \mathbf{X}, \ \mathbf{W} = \mathbf{G}^{-1}\mathbf{X}. \quad (2.41)$$

The stripe location on the camera coordinate is, thus, rectified as

$$x_c' = \frac{W_1 + W_2 x_p}{1 - W_3 x_p} \cdot \quad (2.42)$$

2.4.5 Solution Using the Geometrical Cue

Equation (2.31) can be written as

$$v_c v_p V_1 + v_p x_c V_2 + v_c x_p V_3 - x_p x_c V_4 = 0 \cdot \quad (2.43)$$

For an illumination pattern with n ($n \geq 3$) stripes received on the image plane, (2.39) can be expressed as

$$\begin{bmatrix} v_p v_c & v_c A_1 & v_p X_1 & -A_1 X_1 \\ v_p v_c & v_c A_2 & v_p X_2 & -A_2 X_2 \\ & \cdots & \cdots & \\ v_p v_c & v_c A_n & v_p X_n & -A_n X_n \end{bmatrix}\begin{bmatrix} V_1 \\ V_2 \\ V_3 \\ V_4 \end{bmatrix} = 0, \quad (2.44)$$

$$\text{or } \mathbf{A} \cdot \mathbf{V} = \mathbf{0}, \tag{2.45}$$

where $\mathbf{A}$ is an $n \times 4$ matrix, $X_i = x_c(i)$, $A_i = x_p(i)$, and $\mathbf{V}$ is a 4×1 vector formed from (2.34) to (2.37). The following theorem is used for solving (2.44).

Theorem 2.1 *(The rank of the matrix A) Rank(A) = 3.*

Proof. Consider the 3×3 matrix $\mathbf{A}_{lt}$ in the left-top corner of the $n \times 4$ matrix $\mathbf{A}$. If $\det(\mathbf{A}_{lt}) \neq 0$, then rank($\mathbf{A}$) ≥ 3 is true.

$$\mathbf{A}_{lt} = \begin{bmatrix} v_p v_c & v_c A_1 & v_p X_1 \\ v_p v_c & v_c A_2 & v_p X_2 \\ v_p v_c & v_c A_3 & v_p X_3 \end{bmatrix} \tag{2.46}$$

With row operations, it may be transformed to

$$\mathbf{A}_{lt}' = \begin{bmatrix} v_p v_c & v_c A_1 & v_p X_1 \\ 0 & v_c (A_2 - A_1) & v_p X_2 - v_p X_1 \\ 0 & 0 & v_p (X_3 - X_1) - v_p (X_2 - X_1) \dfrac{A_3 - A_1}{A_2 - A_1} \end{bmatrix}. \tag{2.47}$$

With (2.32) and (2.33), we have

$$U_2 = \frac{(V_2 V_3 - V_1 V_4)}{V_4^2} = \frac{v_c v_p (1 + A_0^2) C_2 (C_1 b + C_2 + h)}{(C_2 + b C_1 + h C_1 A_0)^2} . \tag{2.48}$$

Suppose that the observed line does not pass through the optical center of either the camera or the projector (otherwise it is not possible for triangulation), i.e.

$$C_2 \neq 0, \text{ and } C_1 b + C_2 + h = h - Z(0,b) \neq 0 . \tag{2.49}$$

Hence $U_2 \neq 0$.

For any pair of different light stripes illuminated by the projector, i.e. $A_i \neq A_j$, from (2.32),

$$X_i = U_1 + \frac{U_2}{U_3 + f_p A_i} \tag{2.50}$$

we have $X_i \neq X_j$, and

$$\mathbf{A}_{lt}'(1,1) = v_c v_p \neq 0 , \tag{2.51}$$

$$\mathbf{A}_{lt}'(2,2) = v_c (A_2 - A_1) \neq 0 , \tag{2.52}$$

$$\mathbf{A}_{lt}'(3,3) = v_p(X_3 - X_1) - v_p(X_2 - X_1)\frac{A_3 - A_1}{A_2 - A_1} =$$

$$\frac{v_p U_2 f_p^2 (A_1 - A_3)(A_2 - A_3)}{(U_3 + f_p A_1)(U_3 + f_p A_2)(U_3 + f_p A_3)} \neq 0 \tag{2.53}$$

Hence, $rank(\mathbf{A}) \geq rank(\mathbf{A}_{lt}) = rank(\mathbf{A}_{lt}') = 3$.

On the other hand, rewrite the matrix $\mathbf{A}$ with four column vectors, i.e.

$$\mathbf{A} = [c_{m1} \quad c_{m2} \quad c_{m3} \quad c_{m4}], \tag{2.54}$$

where

$$c_{m1} = [v_p v_c \quad v_p v_c \quad \cdots \quad v_p v_c]^T, \tag{2.55}$$

$$c_{m2} = [v_c A_1 \quad v_c A_2 \quad \cdots \quad v_c A_n]^T, \tag{2.56}$$

$$c_{m3} = [v_p X_1 \quad v_p X_2 \quad \cdots \quad v_p X_n]^T, \tag{2.57}$$

$$c_{m4} = [-X_1 A_1 \quad -X_2 A_2 \quad \cdots \quad -X_n A_n]^T. \tag{2.58}$$

With the fourth column,

$$c_{m4} = \{-X_i A_i\} = \{-\frac{U_2 + U_1 U_3}{f_p} - U_1 A_i + \frac{U_3}{f_p} X_i\}$$

$$= \{\tau_1 v_c v_p + \tau_2 v_c A_i + \tau_3 v_p X_i\} = \tau_1 c_{m1} + \tau_2 c_{m2} + \tau_3 c_{m3}. \tag{2.59}$$

This means that the matrix's 4th column, c_{m4}, has a linear relationship with the first three columns, c_{m1} - c_{m3}. So the maximum rank of matrix $\mathbf{A}$ is 3, i.e. $rank(\mathbf{A}) \leq 3$.

Therefore, we can conclude that $rank(\mathbf{A}) = 3$. φ

Now, considering three pairs of stripe locations on both the camera and the projector, $\{(X_i, A_i) \mid i = 1, 2, 3\}$, (2.44) has a solution in the form of

$$\mathbf{V} = k[v_1 \, v_2 \, v_3 \, v_4]^T, \, k \in R, \tag{2.60}$$

There exists an uncertain parameter k as the rank of matrix $\mathbf{A}$ is lower than its order by 1. Using singular value decomposition to solve the matrix equation (2.44) to find the least eigenvalue, the optimal solution can be obtained. In a practical system setup, the z-axis displacement h is adjusted to 0 during an initial calibration, and (2.60) gives a solution for the relative orientation:

$$\begin{cases} A_0 = \dfrac{v_3}{v_4} \\ b_c = \dfrac{A_0 v_2 - v_1}{A_0 v_1 - v_2} \\ C_1 = (A_0 + b_c)\dfrac{v_4}{v_1} - b_c \end{cases}, \tag{2.61}$$

where $b_c = C_2 / b$.

The orientation of the projector is

$$\alpha_0 = \tan^{-1}(A_0).$$ (2.62)

By setting $b = 1$ and solving (2.33) and (2.60), the 3D reconstruction can be performed to obtain an object shape (with relative size). If we need to obtain the absolute 3D geometry of the object, (2.60) is insufficient for determining the five unknowns, b, C_1, C_2, A_0, and k. To determine all these parameters, at least one more constraint equation is needed. In our previous work (Chen 2003), the focus cue or the best-focused distance is used below for this purpose.

2.5 Summary

Compared with passive vision methods which feature low-cost and are easy to set up, the structured light system using the active lighting features high accuracy and reliability. The stripe light vision system can achieve a precision at the order of 0.1 mm when using a high resolution camera and employing a good sub-pixel method. A specialized projector can be designed for achieving low cost and high accuracy. However, the limitation of the stripe light vision system is that it requires the scene to be of uniform color and static within the acquisition period. To reconstruct one 3D surface, it needs about one second to capture 8–12 images and several seconds to compute the 3D coordinates. The two methods of stereo vision and stripe light vision have both been summarized in this chapter.

To make the vision system flexible for perceiving objects at varying distances and sizes, the relative position between the projector and the camera needs to be changeable, leading to reconfiguration of the system. A self-recalibration method for such a reconfigurable system needs to be developed to determine the relative matrix between the projector and the camera, which is mainly concerned in this chapter. We thus presented a work in automatic calibration of active vision systems via a single view without using any special calibration device or target. This is also in the field of "self-calibration" and "self-recalibration" of active systems. Here self-recalibration deals with situations where the system has been initially calibrated but needs to be calibrated again due to a changed relative pose (the orientation and position) between the camera and projector. Self-calibration refers to cases where the system has never been calibrated and none of the sensor's parameters including the focal length and relative pose are known. Although the method described in this chapter is mainly concerned with the former situation, it can also be applied to the latter case. Some important cues are explored for recalibration using a single view. The method will be applicable in many advanced robotic applications where automated operations entail dynamically reconfigurable sensing and automatic recalibration to be performed on-line without operators' interference.

Chapter 3
Active Sensor Planning – the State-of-the-Art

The aim of sensor planning is to determine the pose and settings of a vision sensor for undertaking a vision task that usually requires multiple views. Planning for robot vision is a complex problem for an active system due to its sensing uncertainty and environmental uncertainty. This chapter describes the problem of active sensor planning formulated from practical applications and the state-of-the-art in this field.

3.1 The Problem

An active visual system is a system which is able to manipulate its visual parameters in a controlled manner in order to extract useful data about the scene in time and space. (Pahlavan et al. 1993)

Active sensor planning endows the observer capable of actively placing the sensor at several viewpoints through a planning strategy. In the computer vision community, when active perception became an important attention to researchers, sensor planning inevitably became a key issue because the vision agent had to decide "where to look". According to task conditions, the problem is classified into two categories, i.e. model-based and non-model-based vision tasks.

About 20 years ago, Bajcsy discussed the important concept of active perception (Bajcsy 1988). Together with other researchers" initial contributions at that time, the new concept (compared with the Marr paradigm in 1982) on active perception, and consequently the sensor planning problem, was thus issued in vision research. The difference between the concepts of active perception and the Marr paradigm is that the former considers vision perception as the intentional action of the mind but the latter considers it as the procedural process of matter.

Therefore, research of sensor planning falls into the area of active perception (Bajcsy 1988). It introduces the idea of moving a sensor to constrain interpretation of its environment. Since multiple 3D images need to be taken and integrated from different vantage points to enable all features of interest to be measured, sensor placement which determines the viewpoints with a viewing strategy thus becomes critically important for achieving full automation and high efficiency.

The problem of sensor placement in computer vision was addressed by Tarabanis et al. (1995) as: "for a given information concerning the environment (object under observation, sensor available) and concerning the task that the system

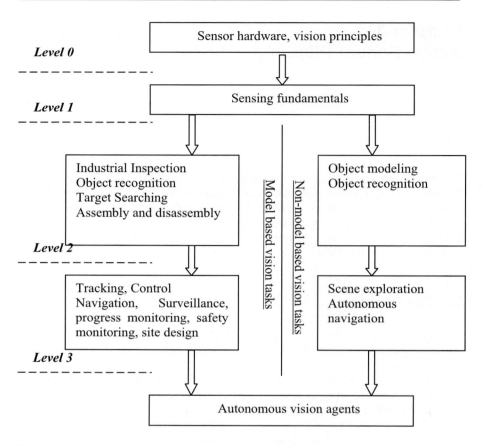

Fig. 3.1. The roles of active sensor planning in autonomous robots

must achieve (detection of characteristics, object recognition, scene reconstruction), to develop some automatic strategy to determine the sensor parameters (the position, the orientation and the optical parameters of the sensor) to carry out the task satisfying some criteria."

Today, the roles of sensor planning can be widely found in most autonomous robotic systems. According to the task conditions, the planning scheme can be applied on different levels of vision perception as illustrated in Fig. 3.1.

3.2 Overview of the Recent Development

The early work on sensor planning was mainly focused on the analysis of placement constraints, such as resolution, focus, field of view, visibility, and conditions for light source placement in 2D space (Lin et al. 1996). A viewpoint has to be placed in an acceptable space and a number of constraints should be satisfied. The fundamentals in solving such a problem were established in the last decades. Tarabanis et al. (1995) presented an intensive survey on sensing strategies developed in the early stage, concentrated upon the period between 1987 and 1991. Among them, Cowan (1988) gave detailed descriptions on computing the acceptable viewpoints for satisfying many requirements (sensor placement constraints). In Cowan (1988), lens aperture setting was also considered by computing the diffraction limit. The light position region was determined to achieve adequate illumination, mathematically through the light path, i.e. surface absorption, diffused reflectance, specular reflectance, and image irradiance. Abrams et al. (1999) also proposed to compute the viewpoints that satisfy the optical constraints, i.e. resolution, focus (depth of field), field-of-view, and detectability. Rosenfeld discussed some techniques and the relationship between object recognition and known or unknown viewpoints (Rosenfeld 1988). More extensive surveys of the early works can be found in Banta (1996), Marchand (1997, 1999), and Kelly et al. (2000).

Here the scope is restricted to recently published approaches to view-pose determination and sensor optical settings in the robotics community. It does not include: attention, gaze control, foveal sensing, hand-eye coordination, autonomous vehicle control, localization, landmarks, qualitative navigation, path following operation, etc., although these are also issues concerning the active perception problem.

Of the published literature in the recent years, Cowan (1988) is one of the earliest research on this problem in 1988 although some primary works can be found in the period 1985–1987. To date, there are more than two hundred research papers which mainly focus on sensor placement or viewpoint planning. At the early stage, these works were focused on sensor modeling, analysis of sensors" optical and geometrical parameters, and sensor placement constraints. From 1990 to 1995, most of these research works were CAD model-based and usually for applications in computer inspection or recognition. The generate-and-test method and the synthesis method are major contributions at that stage. From 1996 to 2000, while optimization was still necessary for model-based sensor placement, it is increasingly important to plan viewpoints for unknown objects or no a priori environment because this is very useful for many active vision tasks such as model reconstruction and autonomous navigation. In recent years, although researchers have to continue working on the theoretical formulation of active sensor planning, many works tend to combine the existing methods with specific application such as inspection, recognition, search, object modeling, tracking, exploration, navigation, localization, assembly and disassembly, etc.

Two outstanding methods have been widely used previously. They are the weighted function and tessellated space. The former uses a function that includes several components standing for placement constraints, e.g.

$$h = \max(\alpha_1 g_1 + \alpha_2 g_2 + \alpha_3 g_3 + \alpha_4 g_4) \tag{3.1}$$

Equivalently with constraint-based space analysis, for each constraint (such as visibility, resolution, field-of-view, and depth-of-field), the sensor pose is limited to a possible region. Then the viewpoint space is the intersection of these regions and the optimization solution is determined by the above function in the viewpoint space, i.e.,

$$V_{\text{placement}} = V_{g1} \cap V_{g2} \cap V_{g3} \cap V_{g4} \tag{3.2}$$

This method is usually used in model-based planning (Trucco 1997) tasks, such as inspection, assembly/disassembly, recognition, and object search.

The latter method tessellates a sphere or cylinder around the object to be modeled as a viewpoint space (or look-up array (Morooka et al. 1999)). Each grid point is a possible sensor pose for viewing the object. The object surface is partitioned as void surface, seen surface, unknown surface, and uncertain surface. The working space is also partitioned into void volume and viewing volume. Finally an algorithm is employed for planning a sequence of viewpoints so that the whole object can be sampled. This method is effective in dealing with some small and simple objects, but it is difficult to model a large and complex object with many concave areas because it cannot solve occlusion constraint.

More precisely, a number of approaches have been applied in deciding the placement of the vision sensor, including:

- geometrical/ volumetric computation
- tessellated sphere/space -TS
- generate-and-test approach (Kececi 1998, Trucco 1997)
- synthesis approach
- sensor simulation
- expert system
- rules (Liu and Lin 1994)
- iterative optimization method (Lehel et al. 1999)
- Bayesian decision (Zhou and Sakane 2001, Kristensen 1997)
- probabilistic reasoning (Roy 2000)
- tree annealing (Yao 1995)
- genetic algorithm (Chen et al. 2004).

Out of these approaches, volumetric computation by region intersection is most frequently used by researchers, e.g. (Cowan 1988). For each constraint, it computes the region Ri of acceptable viewpoints. If multiple surface features need to be inspected simultaneously, the region Ri is the intersection of the acceptable regions Rij for each individual feature. Finally, the region of acceptable viewpoints is the intersection of all regions.

3.3 Fundamentals of Sensor Modeling and Planning

Table 3.1 lists some fundamental works on sensor modeling and vision planning for robotic tasks. It provides an overview of typically used sensors, controllable parameters, proposed methods, and applied tasks.

Table 3.1. Summary of typical works on fundamental sensor planning

Reference	Sensors	Parameters	Method	Task
Cowan 1988	Camera; Extension to laser scanner	Resolution, focus (depth of field), field of view, visibility, view angle; 6 extrinsic parameters of the sensor	Geometrical computation	General model based vision task
Tarabanis 1991	Camera	Optical constraints (resolution, focus/ depth-of-field, field-of-view, and detectability)	Volume intersection method VIM	General purpose
Remagnino1995	Camera	Position, look direction (pan/tilt), focal length	Geometrical computation	General task in partially known environment
Giraud 1995	General sensors	Perception number, sensor location	Geometrical approach, Bayesian statistics	Equipment design, general task
Triggs 1995	Camera	Task, camera, robot and environment	Probabilistic heuristic search, combined evaluation function	General model based vision task
Yao 1995	Camera	Generalized viewpoint, depth of field, field of view, resolution	Tree annealing	General model based vision task
Tarabanis 1995	Camera	Camera pose, optical settings, task constraints	VIM	Model based vision task
Stamos1998	Camera	Field-of-view, visibility	Interactive	General model based vision task
Lehel et al. 1999	Trinocular sensor (CardEye)	Relative intrinsic translation, pan, tilt, field of view angle	Iterative optimization	General vision tasks
Li and Liu 2003	Structured light	Reconfigured pose	Geometrical	Recalibration for active vision
Zanne et al. 2004	Eye-in-hand camera	Path	Constraint-based control	Visibility

(Continued)

Table 3.1. (Continued)

Reference	Sensors	Parameters	Method	Task
Farag 2004	Trinocular	Center, zoom, focus and vergence	SIFT algorithm	Mobile vision system
Mariottini 2005	Pinhole and panoramic cameras	Camera intrinsic and relative parameters	Geometrical modeling	Camera models and epipolar geometry
LaValle 2006	general	NA	Algorithms using information space, differential constraints, etc.	Motion planning
Hua et al. 2007	Panoramic Camera	Wide FOV, high-resolution	Mirror pyramid	Maximize the panoramic FOV

For active sensor planning, an intended view must first satisfy some constraints, either due to the sensor itself, the robot, or its environment. From the work by Cowan et al. (1988) who made a highlight on the sensor placement problem, detailed descriptions of the acceptable viewpoints for satisfying many requirements (sensor placement constraints) have to be provided. Cowan and Kovesi (1988) presented an approach to automatically generating camera locations (viewpoints), which satisfied many requirements (we term it sensor placement constraints) including resolution, in-focus, field-of-view, occlusion, etc. Shortly after that, they (Cowan and Bergman 1989) further described an integrated method to position both a camera and a light source. Besides determining the camera placement region to satisfy the resolution, field of view, focus, and visibility, lens aperture setting was also considered by computing the diffraction limit. The light position region was determined to achieve adequate illumination, mathematically through the light path, i.e. surface absorption, diffused reflectance, specular reflectance, and image irradiance. Similar concepts were also presented by Tarabanis et al. (1991) to compute the viewpoints that satisfy the sensing constraints, i.e. resolution, focus, field-of-view, and detectability. A complete list of constraints will be summarized and analyzed in Chap. 4.

To better describe the sensor properties, Ikeuchi et al. (1991) presented a sensor modeler, called VANTAGE, to place the light sources and cameras for object recognition. It mostly proposed to solve the detectability (visibility) (Zanne et al. 2004) of both light sources and cameras. It determined the illumination/observation directions using a tree-structured representation and AND/OR operations. The sensor is defined as consisting of not only the camera, but multiple components (G-sources), e.g. a photometric stereo. It is represented as a sensor composition tree (SC tree), as in Fig. 3.2. Finally, the appearance of object surfaces is predicted by applying the SC tree to the object and is followed by the action of sensor planning.

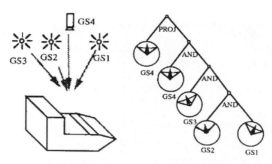

Fig. 3.2. The photometric stereo sensor and its SC tree (Ikeuchi and Robert 1991)

In some other typical works on constraint formulation, Remagnino et al. (1995) proposed to set the viewpoint, look direction, and focal length of a camera. With a partially known environment, it dealt with two problems: how to determine the sensor"s pose (in the bootstrap phase) and how to determine the next-look direction (in the run phase). It took into account errors in the object position stored in the memory and errors due to image segmentation. Rosenfeld et al. (1988) discussed some techniques and relationship between object recognition and known or unknown viewpoints. In fact, an intensive survey on sensing strategies developed in the first stage, i.e. the period from 1987 to 1992, was summarized by Tarabanis et al. (1995).

To a relatively higher level, Giraud and Jouvencel (1995) addressed the sensor selection at an abstract level for equipment design and perception planning. It is formulated with (1) the number of uses of a sensor; (2) the selection of multi-sensors; (3) discarding useless sensors; and (4) the location of the sensors. It used an approach based on geometrical interaction between a sensor and an environment and Bayes reasoning to estimate the achieved information. Later, Kristensen et al. (1997) proposes the sensor planning approach also using the Bayesian decision theory. The sensor modalities, tasks, and modules were described separately and the Bayes decision rule was used to guide the behavior.

The model-based sensor placement problem in fact is formulated as a nonlinear multi-constraint optimization problem. It is difficult to compute robust viewpoints which satisfy all feature detectability constraints. Yao and Allen (1995) presented a tree annealing (TA) method to compute the viewpoints with multi-constraints. They also investigated the stability and robustness while considering the constraints with the different scale factors and noises. Another way is done by Triggs and Laugier (1995) who described a planner to produce heuristically good static viewing positions. It combined many task, camera, robot and environmental constraints. A viewpoint is optimized and evaluated by a function which uses a probability-based global search technique.

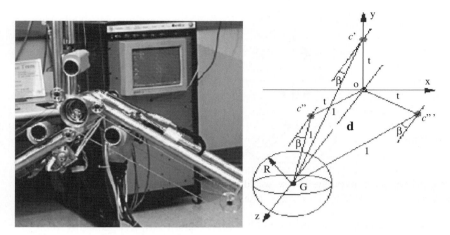

Fig. 3.3. The CardEye trinocular vision sensor and its model (with the Computer Vision and Image Processing Lab (CVIP) at the University of Louisville (Farag 2004))

In a recent book by Steve LaValle (2006), many different kinds of planning algorithms can be found related to visibility and sensor-based planning, e.g. information space, differential constraints, decision-theoretic planning, sampling-based planning, combinatorial planning, etc.

For active sensing purpose, many devices and systems have recently been invented for robotics, e.g. (Colin 2007, Hou et al. 2006). An ATRV-2 based AVENUE mobile robot is used by Blaer and Allen (2006) for automated site modeling (Fig. 3.4), at the Columbia University. Sheng et al. (2006) develop an automated, intelligent inspection system for these engineered structures, which

Fig. 3.4. The ATRV-2 based AVENUE mobile robot for automated site modeling (Blaer and Allen 2006)

employs a team of intelligent climbing robots and a command robot to collaboratively carry out the inspection task. To support autonomous navigation, a miniature active camera (MoCam) module is designed, which can be used in the pose calibration of the robot. Farag (2004) solves the planning problem for a mobile active system with a trinocular vision sensor (Fig. 3.3). An algorithm is proposed to combine a closed-form solution for the translation between the three cameras, the vergence angle of the cameras as well as zoom and focus setting with the results of the correspondences between the acquired images and a predefined target obtained using the Scale Invariant Feature Transform (SIFT) algorithm. There are two goals. The first is to detect the target objects in the navigation field. The second goal is setting the cameras in the best possible position with respect to the target by maximizing the number of correspondences between the target object and the acquired images. The ultimate goal for the algorithm is to maximize the effectiveness of the 3D reconstruction from one frame.

For fast development of sensor modeling, Mariottini and Prattichizzo (2005) develop an Epipolar Geometry Toolbox (EGT) on MATLAB which is a software package targeted to research and education in computer vision and robotic visual servoing (Fig. 3.5). It provides the user with a wide set of functions for designing multicamera systems for both pinhole and panoramic cameras. Several epipolar geometry estimation algorithms have been implemented. They introduce the toolbox in tutorial form, and examples are provided to demonstrate its capabilities. The complete toolbox, detailed manual, and demo examples are freely available on the EGT Web site (http://egt.dii.unisi.it/).

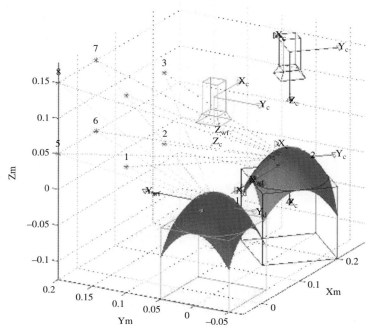

Fig. 3.5. The camera model for visual servoing (Mariottini and Prattichizzo 2005)

3.4 Planning for Dimensional Inspection

In many vision tasks, there exists an object model in the system. For example, in assembly (Nelson 1996), model-based recognition (Okamoto 1998), object searching, dimensional measurement, inspection, and semi-automated scene reconstruction, the object"s geometry and a rough estimate of its pose are known. Especially for the inspection tasks, using either range sensors (Prieto 1999) or intensity cameras (Gu et al. 1999, Abrams 1999), a nearly perfect estimate of the object"s geometry and possibly its pose are known and the task is to determine how accurately the object has been manufactured. Table 3.2 lists some typical works on sensor planning for automated inspection.

On object inspection, Yao and Allen argued that this problem in fact was a nonlinear multi-constraint optimization problem (Yao 1995). Triggs and Laugier (1995) described a planner to produce heuristically good static viewing positions. It combined many task, camera, robot and environmental constraints. A viewpoint is optimized and evaluated by a function which uses a probability-based global search technique. It is difficult to compute robust viewpoints which satisfy all feature detectability constraints. Yao and Allen (1995) presented a Tree Annealing (TA) method to compute the viewpoints with multi-constraints. They also investigated the stability and robustness while considering the constraints with the different

Table 3.2. Summary of typical works on sensor planning for dimensional inspection

Reference	Sensors	Parameters	Method	Task
Tarabanis 1995	Camera	Camera pose, optical settings, task constraints	VIM	Model based vision task
Abrams 1996	Camera	Detectability, in focus, field-of-view, visibility, and resolution	VIM	Inspection
Trucco 1997	Generalized sensor	Visibility, reliability, shortest path	Generate-and-test, VIM, FIR, CCAO	Inspection
Prieto 1999	Range sensor	Viewing distance, incident angle	Direct computation	Inspection
Sheng et al. 2003				
Hodge 2003	Multiple cameras	Positions	Agent-based coordination	Inspection
Chen et al. 2004	Camera, structured light	Camera pose, settings, task constraints	Genetic algorithm, graph theory	Model-based inspection, robot path
Rivera-Rios 2005	Stereo	Camera pose	Probabilistic analysis	Dimensional measurements
Bodor 2005	Cameras	Internal and external camera parameters	Analytical formulation	Observability

scale factors and noises. Elsewise, Olague and Mohr (1998) chose to use genetic algorithms to determine the optimal sensor placements.

In order to obtain a quality control close to the accuracy obtained with a coordinate measuring machine in metrology for automatic inspection, F. Prieto et al. suggest improving the accuracy of the depth measurements by positioning the sensor"s head according to a strategy for optimum 3D data acquisition (Prieto 1999). This strategy guarantees that the viewpoints found meet the best accuracy conditions in the scanning process. The proposed system requires the part"s CAD model to be in IGES format.

Several sensor planning systems have been developed by researchers. For example, Trucco et al. (1997) developed a general automatic sensor planning (GASP) system. Tarbox and Gottschlich (1999) had an Integrated Volumetric Inspection System (IVIS) and proposed three algorithms for inspection planning. Tarabanis et al. (1995) developed a model-based sensor planning system, the machine vision planner (MVP), which works with 2D images obtained from a CCD camera.

Compared with other vision sensor planning systems, the MVP system is notable in that it takes a synthesis rather than a generate-and-test approach, thus giving rise to a powerful characterization of the problem. In addition, the MVP system provides an optimization framework in which constraints can easily be incorporated and combined. The MVP system attempts to detect several features of interest in the environment that are simultaneously visible, inside the field of view, in focus, and magnified, by determining the domain of admissible camera locations, orientations, and optical settings (Fig. 3.6). A viewpoint is sought that is both globally admissible and central to the admissibility domain.

Based on the work on the MVP system (Tarabanis 1995), Abrams et al. (1996) made a further development for planning viewpoints for vision tasks within a robot work-cell. The computed viewpoints met several constraints such as detectability, in-focus, field-of-view, visibility, and resolution. The proposed viewpoint computation algorithm also fell into the "volume intersection method" (VIM). The planning procedure was summarized as: (1) Compute the visibility volumes Vivis; (2) compute the volumes ViFR combined with field-of-view and resolution constraints; (3) compute the overall candidate volume Vc as the intersection of all ViFR and Vivis; (4) find a position within Vc; (5) find the orientation; (6) compute the focus and maximum aperture; (7) verify that the parameters are all valid.

These is generally a straightforward but very useful idea. Many of the latest implemented planning systems can be traced back to this contribution. For example, Rivera-Rios et al. (2005) presents a probabilistic analysis of the effect of the localization errors on the dimensional measurements of the line entities for a parallel stereo setup (Fig. 3.7). The probability that the measurement error is within an acceptable tolerance was formulated as the selection criterion for camera poses. The camera poses were obtained via a nonlinear program that minimizes the total mean square error of the length measurements while satisfying the sensor constraints.

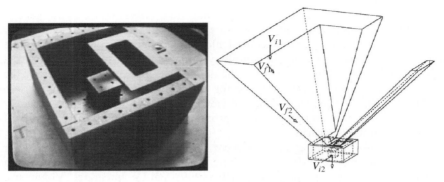

Fig. 3.6. The admissible domain of viewpoints (Tarabanis 1995)

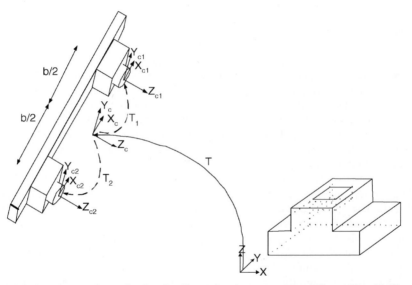

Fig. 3.7. Stereo pose determination for dimensional measurement (Rivera-Rios 2005)

The general automatic sensor planning system (GASP) reported by Trucco et al. (1997) is to compute optimal positions for inspection tasks using a known imaging sensor and feature-based object models. This exploits a feature inspection representation (FIR) which outputs an explicit solution off-line for the sensor position problem. A generalized sensor (GS) model was defined with both the physical sensor and the particular software module. The viewpoints are planned by computing the visibility and reliability. The reliability of the inspection depends on the physical sensors used and the processing software. In order to find a shortest path through the viewpoints in space, they used the Convex hull, Cheapest insertion, angle selection, Or-optimization (CCAO) as the algorithm to solve the traveling salesman problem (Fig. 3.8).

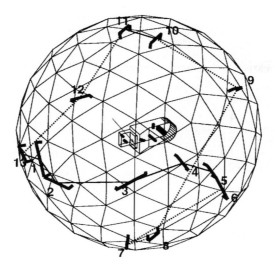

Fig. 3.8. The shortest path planned to take a stereo pair through the viewpoints (Trucco 1997)

In order to obtain a more complete and accurate 3D image of an object, Prieto et al. (1999) presented an automated acquisition planning strategy utilizing its CAD model in IGES format. The work was focused on improving the accuracy of the 3D measured points which is a function of the distance to the object surface and of the laser beam incident angle.

3.5 Planning for Recognition and Search

In many cases, a single view of an object may not contain sufficient features to recognize it unambiguously. Therefore another important application of sensor planning is active object recognition (AOR) which recently attracts much attention within the computer vision community. Object search is also considered a model-based vision task concerned with finding a given object in a known or unknown environment. The object search task not only needs to perform the object recognition and localization, but also involves sensing control, environment modeling, and path planning. Sensor planning is very important for 3D object search since a robot needs to interact intelligently and effectively with the 3D environment. Table 3.3 lists the typical works on sensor planning for vision-based recognition and search.

Table 3.3. Summary of typical works on sensor planning for recognition and search

Reference	Sensors	Parameters	Method	Task
Ikeuchi and Robert 1991	Light source, camera	Illumination/ observation directions	Tree-structured, logical operation	Object recognition
Ye 1995	Camera + range finder	Sensing pose, search space	Probability, tessellated sphere (TS)	Object search
Liu and Lin 1994 Lin et al. 1996	Structured light	View pose	Rules, feature prediction, MERHR	Recognition
Madsen and Christensen 1997	Camera	Viewing direction	Direct computation	Active object recognition (AOR)
Borotschnig 2000	Camera, illuminant	View pose	Probabilistic object classifications, score ranking	AOR
Deinzer 2000	Camera	Classification and localization	Reinforcement learning	AOR
Roy 2000	Camera	View pose, object features	Probabilistic reasoning, Bayes rule	AOR
Sarmiento et al. 2005	General sensor	Sensing locations	Convex cover algorithm	Object search
Xiao et al. 2006	Sonar and omni-directional camera	Path	Fuzzy logic algorithm	Search

In fact, two objects may have all views in common with respect to a given feature set, and may be distinguished only through a sequence of views (Roy 2000). Further, in recognizing 3D objects from a single view, recognition systems often use complex feature sets. Sometimes, it may be possible to achieve the same, incurring less error and smaller processing cost by using a simpler feature set and suitably planning multiple observations. A simple feature set is applicable for a larger class of objects than a model base with a specific complex feature set. Model base-specific complex features such as 3D invariants have been proposed only for special cases. The purpose of AOR is to investigate the use of suitably planned multiple views for 3D object recognition. Hence the AOR system should also take a decision on "where to look". The system developed for this task is an iterative active perception system that executes the acquisition of several views of the object, builds a stochastic 3D model of the object and decides the best next view to be acquired. Okamoto et al. (1998) proposed such a method based on an entropy measure. Liu and Lin (1994), Lin et al. (1996), and Madsen and Christensen (1997) proposed their sensor planning strategies for recognition using rules to automatically predict and detect object features and calculate the next sensor pose, and they applied the maximum expected rate of hypothesis reduction (MERHR) to minimize the sensing actions. Madsen and Christensen's strategy (Madsen and Christensen 1997) was to determine the true angle on the object surface. It automatically guided a movable camera to a position where the optical axis is perpendicular to a plane spanned by any two intersecting edges on a polyhedral object, so that it could determine the true angle of a junction and align the camera. Ye and Tsotsos (1999) used a strategy for object search by planning the sensing actions on the sensed sphere or layered sensed sphere. It was based on a mobile platform, an ARK robot, equipped with a Laser Eye with pan and tilt capabilities. They combined the object recognition algorithm and the target distribution probability for the vision task.

Ikeuchi et al. (1991) developed a sensor modeler, called VANTAGE, to place the light sources and cameras for object recognition. It mostly solves the detectability (visibility) of both light sources and cameras. Borotschnig et al. (2000) also presented an active vision system for recognizing objects which are ambiguous from certain viewpoints. The system repositions the camera to capture additional views and uses probabilistic object classifications to perform view planning. Multiple observations lead to a significant increase in recognition rate. The view planning consists in attributing a score to each possible movement of the camera. The movement obtaining the highest score will be selected next (Fig. 3.9). It was based on the expected reduction in Shannon entropy over object hypotheses given a new viewpoint, which should consist in attributing a score $s_n(\Delta\psi)$ to each possible movement $\Delta\psi$ of the camera. The movement obtaining the highest score will be selected next:

$$\Delta\psi_{n+1} := \arg\max s_n(\Delta\psi) \tag{3.3}$$

Reinforcement learning has been attempted by Deinzer et al. (2000) for viewpoint selection for active object recognition and for choosing optimal next views for improving the classification and localization results. Roy et al. (2000) attempted probabilistic reasoning for recognition of an isolated 3D object. Both the probability calculations and the next view planning have the advantage that the

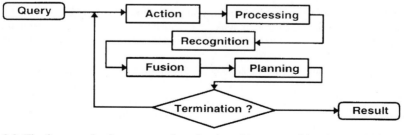

Fig. 3.9. The framework of appearance-based active object recognition (Borotschnig 2000)

knowledge representation scheme encodes feature-based information about objects as well as the uncertainty in the recognition process. The probability of a class (a set of aspects, equivalent with respect to a feature set) was obtained from the Bayes rule (Roy 2000):

$$P(B_i|E^k) = \frac{P(B_i)P(E^k|B_i)}{\sum_j P(B_j)P(E^k|B_j)} \tag{3.4}$$

where $P(B_i \mid E^k)$ is the post-probability of the given subtask done by the action agent.

In the next view planning, two possible moves may be followed from one view to another, i.e. primary move and auxiliary move. A primary move represents a move from an aspect, the minimum angle needed to move out of it. An auxiliary move represents a move from an aspect by an angle corresponding to the primary move of another competing aspect.

3.6 Planning for Exploration, Navigation, and Tracking

On sensor planning for exploration, navigation, and tracking, there is a similar situation that the robot has to work in a dynamic environment and the sensing process may associate with many noises or uncertainties. This issue has become the most active for many applications in recent years. For example, Bhattacharya et al. (2007), Gutmann et al. (2005), Kim (2004), Parker et al. (2004), Steinhaus et al. (2004), Giesler (2004), Yamaguchi et al. (2004), Wong and Jarvis (2004), and Bekris et al. (2004) are related to sensor planning for navigation; Yang et al. (2007), Deng et al. (2005), Chivilo et al. (2004), Harville and Dalong (2004), Thompson (2003), Nishiwaki (2003), and Saeedi et al. (2006) are related to sensor planning for tracking; Huwedi (2006), Leung and Al-Jumaily (2004), and Isler (2003) are related to sensor planning for exploration; Reitinger et al. (2007), Blaer (2006), Ikeda (2006), Park (2003), and Kagami (2003) are related to sensor planning for modeling; and Lim (2003) is for surveillance. Table 3.4 lists the typical works on sensor planning for these topics.

Table 3.4. Some typical works on sensor planning for navigation and modeling

Reference	Sensors	Parameters	Method	Task
Remagnino 1995	Camera	Position, look direction (pan/tilt), focal length	Direct computation	General vision task in partially known environment
Kristensen 1997	General sensor/ actuator	Sensor actions	Bayesian decision	Autonomous navigation in partly known environments
Gracias 2003			Mosaic-based	Underwater navigation

Table 3.4. (Continued)

Reference	Sensors	Parameters	Method	Task
Zhu 2004	Panoramic stereo	Position, orientation	Adaptive	Tracking and localization
Chen et al. 2005	General	Sensor pose	Trend surface	Object modeling
Skrzypc-zynski 2005	Cameras	Position	Landmarks	Positioning, navigation
Murrieta-Cid 2005	Range sensor	Visibility, distance, speed	Differential, system model	Surveillance; maintaining visibility
Hughes and Lewis 2005	Cameras	Camera placement, field of view	Simulation	Exploration
Belkhouche and Belkhouche 2005	General	Robot position and orientation	Guidance laws	Tracking, navigation
Kitamura 2006	Camera, other sensor	Human intervention	Biologically inspired, learning	Navigation
Ludington et al. 2006	Aerial camera	Position	Vision-aided inertial, probability	Navigation, search, tracking
Bhattacharya et al. 2007	Camera	Path, field of view, camera pan	Region based	Landmark-based navigation

For navigation in an active way, a robot is usually equipped with a "controllable" vision head, e.g. a stereo camera on pan/tilt mount (Fig. 3.10). Kristensen (1997) presented the problem of autonomous navigation in partly known environments (Fig. 3.11). Bayesian decision theory was adopted in the sensor planning approach. The sensor modalities, tasks, and modules were described separately and Bayes decision rule was used to guide the behavior. The decision problem for one sensor was constructed with a standard tree for myopic

Fig. 3.10. The robot with an active stereo head (e.g. rotation, pan/tilt mount) (Parker et al. 2004)

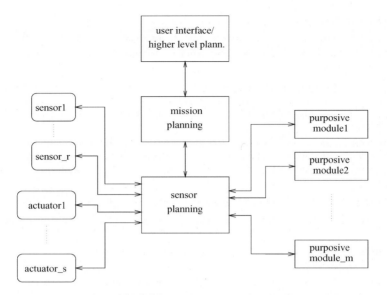

Fig. 3.11. The planning architecture with three levels of abstraction, illustrating that the planner mediates the sensors to the purposive modules (Kristensen 1997)

decision. Object search is also a model-based vision task which is to find a given object in a known or unknown environment. The object search task not only needs to perform object recognition and localization, but also involves sensing control, environment modeling, and path planning.

Zhuang et al. (2004) developed an adaptive panoramic stereovision approach for localizing 3D moving objects at the department of computer science at the University of Massachusetts at Amherst. The research focuses on cooperative robots involving cameras (residing on different mobile platforms) that can be dynamically composed into a virtual stereovision system with a flexible baseline in order to detect, track, and localize moving human subjects in an unknown indoor environment. It promises an effective way to solve the problems of limited resources, view planning, occlusion, and motion detection of movable robotic platforms. Theoretically, two interesting conclusions are given:

1. If the distance from the main camera to the target, D_1, is significantly greater (e.g., ten times greater) than the size of the robot (R), the best geometric configuration is

$$B \approx 2\sqrt{D_1 R}, \ \cos\phi_1 = \frac{3BD_1}{2D_1^2 + B^2} \tag{3.5}$$

where B is the best baseline distance for minimum distance error and ϕ_1 is the main camera"s inner angle of the triangle formed by the two robots and the target.

2. The depth error of the adaptive stereovision is proportional to 1.5 the power of the camera-target distance $(D^{1.5})$, which is better than the case of the best possible fixed baseline stereo in which depth error is proportional to the square of the distance (D^2).

On the visual tracking problem, Belkhouche and Belkhouche (2005) pointed out that the traditional control algorithms based on artificial vision suffered from two problems:

1. The control algorithm has to process in real time a huge flow of data coming from the camera. This task may be difficult, especially for fast tracking problems. Thus, the maximum computational power for image processing is an important issue.
2. The target (or the lead car) is detected only when it appears in the camera"s field of view. Thus, the target must stay in the camera scope of the pursuer. This requirement is necessary to implement a vision-based algorithm.

Therefore, they make a mathematical formulation for modeling and controlling a convoy of wheeled mobile robots. The approach is based on guidance laws strategies, where the robotic convoy is modeled in terms of the relative velocities of each lead robot with respect to its following robot. This approach results in important simplifications to the sensory system as compared to artificial vision algorithms.

Concerning the surveillance problem, there is a decision problem which corresponds to answering the question: can the target escape the observer"s view? Murrieta-Cid et al. (2005) defined this problem and considered to maintain surveillance of a moving target by a nonholonomic mobile observer. The observer"s goal is to maintain visibility of the target from a predefined, fixed distance. The target escapes if

(a) it moves behind an obstacle to occlude the observer"s view,
(b) it causes the observer to collide with an obstacle, or
(c) it exploits the nonholonomic constraints on the observer"s motion to increase its distance from the observer beyond the surveillance distance.

An expression derived for the target velocities is:

$$\begin{pmatrix} \dot{x}_T(t) \\ \dot{y}_T(t) \end{pmatrix} = \begin{pmatrix} \cos\theta & -l\cos\phi \\ \sin\theta & l\cos\phi \end{pmatrix} \begin{pmatrix} u_1 \\ u_3 \end{pmatrix} \qquad (3.6)$$

where θ and ϕ are the observer"s orientation, u_1 and u_3 are moving speeds, and l is the predefined surveillance distance.

To maintain the fixed required distance between the target and the observer, the relationship between the velocity of the target and the linear velocity of the observer is

$$f(u_1, u_3) = u_1^2 + 2u_1 u_3 l \sin(\theta - \phi) + l^2 u_3^2 = 1 \qquad (3.7)$$

Equation (3.7) defines an ellipse in the u_1–u_3 plane and the constraint on u_1 and u_3 is that they should be inside the ellipse while supposing $\dot{x}_T^2 + \dot{y}_T^2 \leq 1$. They deal specifically with the situation in which the only constraint on the target"s velocity is a bound on speed (i.e., there are no nonholonomic constraints on the target"s motion), and the observer is a nonholonomic, differential drive system having bounded speed. The system model is developed to derive a lower bound for the required observer speed. It"s also considered the effect of obstacles on the observer"s ability to successfully track the target.

Biologically inspired, Kitamura and Nishino (2006) use a consciousness-based architecture (CBA) for the remote control of an autonomous robot as a substitute for a rat. CBA is a developmental hierarchy model of the relationship between consciousness and behavior, including a training algorithm (Fig. 3.12). This training algorithm computes a shortcut path to a goal using a cognitive map created on the basis of behavior obstructions during a single successful trial. However, failures in reaching the goal due to errors of the vision and dead reckoning sensors require human intervention to improve autonomous navigation. A human operator remotely intervenes in autonomous behaviors in two ways: low-level intervention in reflexive actions and high-level ones in the cognitive map.

A survey has recently been carried out by Jia et al. (2006). It summarizes the developments of the last 10 years in the area of vision-based target tracking for autonomous vehicle navigation. It concludes that it is very necessary to develop robust visual target tracking based navigation algorithms for the broad applications of autonomous vehicles. Including the recent techniques in vision-based tracking and navigation, some trends of using data fusion for visual target tracking are also discussed. It is especially pointed out that through data fusion the tracking performance is improved and becomes more robust.

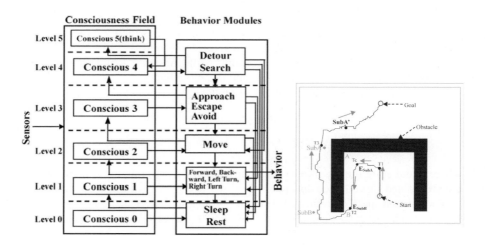

Fig. 3.12. The structure of six-layered consciousness-based architecture and an example of behavior track with intervention (right side)

3.7 Planning for Assembly and Disassembly

For the assembly/disassembly tasks (Table 3.5), a long-term aim in robot programming is the automation of the complete process chain, i.e. from planning to execution. One challenge is to provide solutions which are able to deal with position uncertainties (Thomas et al. 2007, Fig. 3.13). Nelson et al. (1996) introduced a dynamic sensor planning method. They used an eye-in-hand system and considered the resolution, field-of-view, depth-of-view, occlusions, and kinematic singularities. A controller was proposed to combine all the constraints into a system and resulted in a control law. Kececi et al. (1998) employed an independently mobile camera with a 6-DOF robot to monitor a disassembly process so that it can be planned. A number of candidate view-poses are being generated and subsequently evaluated to determine an optimal view pose. A good view-pose is defined with the criterion which prevents possible collisions, minimizes mutual occlusions, keeps all pursued objects within the field-of-view, and reduces uncertainties.

Takamatsu et al. (2002) developed an "assembly-plan-from-observation" (APO) system. The goal of the APO system is to enable people to design and develop a robot that can perform assembly tasks by observing how humans perform those tasks. Methods of contact relations configuration space (C-space) are used to clean up observation errors. Stemmer et al. (2006) use a vision sensor, with color segmentation and affine invariant feature classification, to provide the position estimation within the region of attraction (ROA) of a compliance-based assembly strategy. An assembly planning toolbox is based on a theoretical analysis and the maximization of the ROA. This guarantees the local convergence of the assembly process under consideration of the geometry in part. The convergence analysis uses the passivity properties of the robot and the environment.

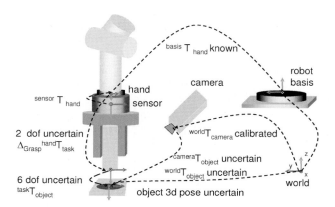

Fig. 3.13. Vision sensor for solving object poses and uncertainties in the assembly work cell (Thomas et al. 2007)

Table 3.5. Some typical works on sensor planning for assembly and disassembly

Reference	Sensors	Parameters	Method	Task
Nelson 1996	Camera	Resolution, FOV, depth-of-view, occlusions, kinematics	Controller (dynamic control law)	Assembly
Kececi 1998	Camera	FOV, view pose, occlusion, uncertainties	Generate-and-test, view-pose assessment/ evaluation	Disassembly
Molineros 2001	Camera	Position	Appearance-based	Assembly planning
Takamatsu 2002	General	Spatial relation	C-space	Assembly, recognition
Hamdi and Ferreira 2004	Virtual	Position	Physical-based	Microassembly
Kelsey et al. 2006	Stereo	Pose	Model-based, synthetic	Pose estimation and tracking
Thomas et al. 2007	Cameras	Relative poses	Multi sensor fusion	Assembly

3.8 Planning with Illumination

Table 3.6. Summary of typical works on sensor planning with illumination

Reference	Sensors	Parameters	Method	Task
Cowan 1989	Camera, light source	Camera, light position region	Illumination computation via reflectance	General model based tasks
Ikeuchi and Robert 1991	Light source, camera	Illumination/ observation directions	Tree-structured, logical operation	Object recognition
Eltoft 1995				Enhancing image features
Solomon 1995	Light source, camera	Positions	Model-based	Lambertian polyhedral objects
Racky and Pandit 1999	Light source	Position	Physics	Segmentation
Xu and Zhang 2001	Light source	Pose, intensity, and distribution of light sources	Neural-network	Surgical applications; general vision tasks

Table 3.6. (Continued)

Reference	Sensors	Parameters	Method	Task
Qu 2003				
Spence 2006	Photometric stereo	Position	Sensitivity analysis	Surface measurement
Yang and Welch 2006	Light source	Illumination variance	Illumination estimation	Tracking
Chen et al. 2007	Light source	Intensity, glares	PID-controller	General tasks
Marchand 2007	Light, camera	Positions	Brightness, contrast	Visual servoing

The light source for a natural scene is its illumination. For many machine-vision applications, illumination now becomes the most challenging part of system design, and is a major factor when it comes to implementing color inspection. Table 3.6 lists the typical works on sensor planning with illumination, recently carried out in the robot vision community. Here, when illumination is also considered, the term "sensor" has a border meaning "sensor/actuator/illuminant".

Eltoft and deFigueiredo (1995) found that illumination control could be used as a means of enhancing image features. Such features are points, edges, and shading patterns, which provide important cues for the interpretation of an image of a scene and the recognition of objects present in it. Based on approximate expressions for the reflectance map of Lambertian and general surfaces, a rigorous discussion on how intensity gradients are dependent on the direction of the light is presented. Subsequently, three criteria for the illumination of convex-shaped cylindrical surfaces are given. Two of these, the contrast equalization criterion and the max-min equalization criterion, are developed for optimal illumination of convex polyhedrons. The third, denoted shading enhancement, is applicable for the illumination of convex curved objects. Examples illustrate the merit of the criteria presented

Xu and Zhang (2001) and Zhang (1998) apply a method of modeling human strategy in controlling a light source in a dynamic environment to avoid a shadow and maintain appropriate illumination conditions. Ikeuchi et al. (1991) investigate the illumination conditions with logical operations of illuminated regions. Their developed sensor modeler, VANTAGE, determines the illumination directions using a tree-structured representation and AND/OR operations (Fig. 3.14).

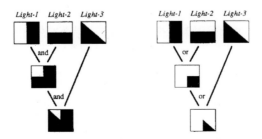

Fig. 3.14. Set operations ("AND" and "OR") among illuminated regions (Ikeuchi and Robert 1991)

Qu et al. (2003) discussed that irradiance distribution and intensity of the test object play a key role in accuracy and stability of the vision measuring system. They proposed a luminance transfer function to design the illumination so that it could adjust light radiation automatically by ways of Neural Networks and Pulse-Width Modulation switch power. They concluded that the illumination could greatly improve the accuracy and robustness of the vision measuring system.

Marchand et al. (2007) recently proposed an approach to control camera position and/or lighting conditions in an environment using image gradient information. The goal is to ensure a good viewing condition and good illumination of an object to perform vision-based tasks such as recognition and tracking. Within the visual servoing framework, the solution is to maximize the brightness of the scene and maximize the contrast in the image. They consider arbitrary combinations of either static or moving lights and cameras. The method is independent of the structure, color and aspect of the objects. For examples, illuminating the Venus of Milo is planned as in Fig. 3.15.

With regard to the placement of the illumination vectors for photometric stereo, Drbohlav and Chantler (2005) discussed the problem of optimal light configurations in the presence of camera noise. Solomon and Ikeuchi proposed an illumination planner for Lambertian polyhedral objects. Spence and Chantler (2006) also found the optimal difference between tilt angles of successive illumination vectors to be 120°. Such a configuration is therefore to be recommended for use with 3-image photometric stereo. Ignoring shadowing, the optimal slant angle was found to be 90° for smooth surfaces and 55° for rough surfaces. The slant angle selection therefore depends on the surface type.

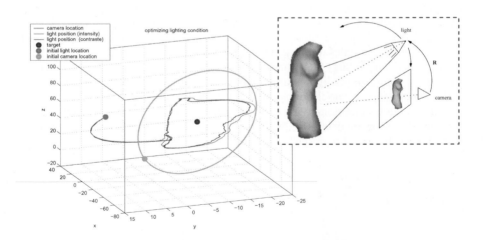

Fig. 3.15. En example of camera and light source position control

3.9 Other Planning Tasks

Besides the tasks already presented in this chapter, there are some other interesting works related to active sensor planning (Table 3.7). For example, Navarro-Serment et al. (2004) describe a method for observing maneuvering targets using a group of mobile robots equipped with video cameras. The cameras seek to observe the target while facing it as much as possible from their respective viewpoints. The work considers the problem of scheduling and maneuvering the cameras based on the evaluation of their current positions in terms of how well can they maintain a frontal view of the target. Some contributions such as interactive planning, virtual placement, robot localization, attention and gaze are briefly introduced below.

3.9.1 Interactive Sensor Planning

In cluttered and complex environments such as urban scenes, it can be very difficult to determine where a sensor should be placed to view multiple objects and regions of interest. Based on their earlier sensor planning results (Tarabanis 1995, Abrams 1999), Stamos and Allen (1998) and Blaer and Allen (2006) extended to build an interactive sensor planning system that can be used to select viewpoints subject to camera visibility, field of view and task constraints. Given a description of the sensor"s characteristics, the objects in the 3D scene, and the targets to be viewed, the algorithms compute the set of admissible view points that satisfy the constraints. The system first builds topologically correct solid models of the scene from a variety of data sources. Viewing targets are then selected, and visibility volumes and field of view cones are computed and intersected to create viewing volumes where cameras can be placed. The user can interactively manipulate the scene and select multiple target features to be viewed by a camera. VRML graphic models and then solid CAD models are assumed as the site models of the scenes (Fig. 3.16).

Table 3.7. Some other interesting works related to active sensor planning

Reference	Sensors	Parameters	Method	Task
Stamos 1998	Camera	Visibility, FOV, task constraints	Interactive	General purpose
Navarro-Serment 2004	Cameras	Positions	Evaluation function	Observation
Zingaretti 2006	Cameras	Relative intrinsic translation, pan, tilt, field of view angle	Partially observable Markov decision	Self-localization
State 2006	Cameras	Visibility, overlap, resolution	Simulation	3D reconstruction in VR
Lidoris et al. 2006	Camera	Gaze direction	Information gain	SLAM

Fig. 3.16. The scene model in which the user can interactively select the target for sensor planning (Stamos and Allen 1998)

With similar tasks, a city model was generated from an incomplete graphics model of Rosslyn VA and was translated by the system to a valid solid model which the planner can use. Overlaid on the city model are the viewing volumes generated for different viewpoints on a selected target face in the scene. The object models and targets can be interactively manipulated while camera positions and parameters are selected to generate synthesized images of the targets that encode the viewing constraints. They extended this system to include resolution constraints (Tarabanis 1995, Allen and Leggett 1995, Reed et al. 1997, Stamos 1998, Abrams 1999).

3.9.2 Placement for Virtual Reality

Interactive camera planning is sometimes also used for virtual reality or simulation. Typical examples can be found from Williams and Lee (2006) and State et al. (2006). For example, the work by State et al. is to simulate in real time multi-camera imaging configurations in complex geometric environments. The interactive visibility simulator helps to assess in advance conditions such as visibility, overlap between cameras, absence of coverage and imaging resolution everywhere on the surfaces of a pre-modeled, approximate geometric dataset of the actual real-world environment the cameras are to be deployed in. A simulation technique is applied to a task involving real-time 3D reconstruction of a medical procedure. It has proved useful in designing and building the multi-camera acquisition system as well as a remote viewing station for the reconstructed data. The visibility simulator is a planning aid requiring a skilled human system designer to interactively steer a simulated multi-camera configuration towards an improved solution.

3.9.3 Robot Localization

As a problem of determining the position of a robot, localization has been recognized as one of the most fundamental problems in mobile robotics. The aim of localization is to estimate the position of a robot in its environment, given local sensorial data. Zingaretti and Frontoni (2006) present an efficient metric for appearance-based robot localization. This metric is integrated in a framework that uses a partially observable Markov decision process as position evaluator, thus allowing good results even in partially explored environments and in highly perceptually aliased indoor scenarios. More details of this topic are related to the research on simultaneous localization and mapping (SLAM) which is also a challenging problem and has been widely investigated (Eustice et al. 2006, Ohno et al. 2006, Lidoris et al. 2006, Herath et al. 2006, Zhenhe and Samarabandu 2005, Jose and Adams 2004, Takezawa et al. 2004, Prasser 2003).

3.9.4 Attention and Gaze

The general concept of active sensor planning should include attention and gaze. This book, however, does not place much emphasis on this issue. Some related works can be found from Bjorkman and Kragic (2004) and (Lidoris et al. 2006). Especially, Bjorkman et al. introduce a real-time vision system that integrates a number of algorithms using monocular and binocular cues to achieve robustness in realistic settings, for tasks such as object recognition, tracking and pose estimation (Fig. 3.17). The system consists of two sets of binocular cameras; a peripheral set for disparity-based attention and a foveal one for higher-level processes. Thus the conflicting requirements of a wide field of view and high resolution can be overcome. One important property of the system is that the step from task specification through object recognition to pose estimation is completely automatic, combining both appearance and geometric models. Experimental evaluation is performed in a realistic indoor environment with occlusions, clutter, changing lighting and background conditions.

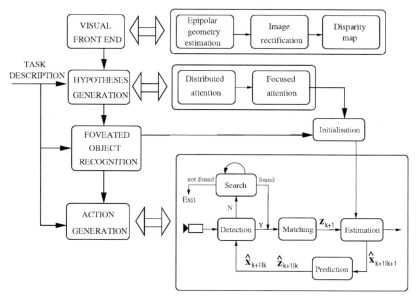

Fig. 3.17. The active vision system involving attention and gaze for action decision (Bjorkman and Kragic 2004)

3.10 Summary

This chapter summarizes the recent development related to the active sensor planning problem. Typical works for inspection, recognition, search, exploration, navigation, tracking, assembly, and disassembly are listed for readers to have a general overview of the state-of-the art.

In model-based tasks, the viewpoint planning is to find a set of admissible viewpoints in the acceptable space, which satisfy a set of the sensor placement constraints and can well finish the vision task. However, the previous approaches are normally formulated for a particular application and are therefore difficult to apply to general tasks. They mainly focus on modeling of sensor constraints and calculating a "good" viewpoint to observe one or several features on the object. Little consideration is given to the overall efficiency of a generated plan with a sequence of viewpoints. However, this method is difficult to apply in a multi-feature-multi-viewpoint problem as it cannot determine the minimum number of viewpoints and their relative distribution.

Therefore a critical problem is still not well solved: the global optimization of sensor planning. When multiple features need to be observed and multiple viewpoints need to be planned, the minimum number of viewpoints needs to be determined. To achieve high efficiency and quality, the optimal spatial distribution of the viewpoints should be determined too. These are also related to the sensor configuration and environmental constraints. Furthermore, to make it flexible in practical applications, we need to deal with arbitrary object models without assumptions on the object features. These problems will be discussed in the following chapters.

Chapter 4
Sensing Constraints and Evaluation

The aim of sensor placement is to determine the pose and settings of a vision sensor for undertaking a vision task that usually requires multiple views. Planning sensor placements is a complex problem for an active visual system. To make each viewpoint practically feasible, a number of constraints have to be satisfied. An evaluation criterion should also be established for achieving an optimal placement plan. This chapter and the next chapter will be concerned with these problems and introduce the corresponding planning strategies.

4.1 Representation of Vision Sensors

The sensor model embodies the information that characterizes the operation of the sensor (Tarabanis 1995). In robotics and manufacturing applications, structured light and stereo systems are widely used. This chapter is concerned with the use of a structured light sensor for accurate acquisition of 3D surface information, whilst the developed method may also be extended to other sensors.

Basically, a sensor has six spatial parameters, i.e. the three positional degrees of freedom of the sensor, $\mathbf{X} = (x, y, z)$, and the three orientational degrees of freedom of the sensor (the pan, tilt, and swing angles), $\mathbf{\Omega} = (\alpha, \beta, \gamma)$. For a stereo sensor, with the assumption that the two cameras are parallel and the baseline is fixed, the sensor's state (viewpoint) may be modeled as a nine-dimensional vector:

$$v = (x, y, z, \alpha, \beta, \gamma, d, f, a),$$

where $\mathbf{\Gamma} = (d, f, a)$ is the optical parameters, including the back principal point to the image plane distance, d; the entrance pupil diameter, a of the lens; and the focal length f of the lens.

For a general 3D range sensor that operates according to the principle of triangulation, i.e. emission of a beam of light (incidental ray) followed by the analysis of the reflected ray, a viewpoint may be similarly defined as a vector of seven parameters (Prieto 1999):

$$v = (x, y, z, \alpha, \beta, \gamma, \psi),$$

where the parameter ψ specifies the angle of the controlled sweep.

For a structured light system as described in a previous chapter, a viewpoint may be defined as a vector of 9 parameters:

$$v = (x, y, z, \alpha, \beta, \gamma, b, \psi, a),$$

where b is the baseline between the camera and the projector. The parameter ψ specifies the relative angle between them, and the parameter a represents the illumination brightness of light projection.

Sensor planning needs to search in such a high-dimensional space (e.g. 9D for structured light systems). A point in that space is defined as a "generalized viewpoint".

4.2 Placement Constraints

While planning a viewpoint to observe an object, the parameters of the sensor position, orientation, and other settings need to be determined. The viewpoint should be planned at an optimal place where it is feasible in the practical environment. This section discusses the placement constraints considered for generating a feasible viewpoint. Usually these constraints include: (1) visibility, (2) resolution, (3) viewing distance or in-focus, (4) field of view, (5) overlap, (6) viewing angle, (7) occlusion, (8) kinematic reachability of sensor pose, (9) robot-environment collision, (10) operation time, etc. The objective of sensor placement is to generate a plan which satisfies all the constraints and has the lowest operation cost. The satisfaction of the placement conditions to constrain the sensor to being placed in the acceptable space is formulated below.

4.2.1 Visibility

The visibility constraint dictates that the admissible sensor pose is limited to regions from where the object surface points to be observed are visible and not occluded. Convex and concave freeform objects with or without holes usually have a visibility problem. A surface patch is said to be visible if the dot product of its normal and the sensor's viewing direction is below 0 (Fig. 4.1a)

$$G1: \quad \mathbf{n}\cdot\mathbf{v}_a = \| \mathbf{n} \| \| \mathbf{v}_a \| \cos(180-\theta) < 0, \ (0<=\theta<=180). \quad (4.1)$$

That gives $\theta < 90°$. To compute the normal direction of the viewpoints, consider a point $A(x, y)$ on the parametric surface S. The normal direction is computed using (Prieto 1999):

$$n = \frac{\frac{\partial}{\partial x}\bar{s}(x,y) \times \frac{\partial}{\partial y}\bar{s}(x,y)}{\left\| \frac{\partial}{\partial x}\bar{s}(x,y) \times \frac{\partial}{\partial y}\bar{s}(x,y) \right\|^2} = \frac{1}{\sqrt{1+p^2+q^2}} \begin{bmatrix} -p \\ -q \\ 1 \end{bmatrix}$$

where the pair $[p, q]$ is a two-dimensional gradient space representation of the surface orientation

$$[p, q]=[\ \frac{\partial z}{\partial x},\ \frac{\partial z}{\partial y}\ ].$$

4.2.2 Viewing Angle

A point is visible if the angle (θ) is less than 90°. However, we should set a limit (θ_{max}) for this angle as the sampling will not be reliable when it is close to 90°, i.e.

G2: $\qquad \theta = \pi - \cos^{-1} \dfrac{\mathbf{n} \cdot \mathbf{v}_a}{\| \mathbf{n} \| \times \| \mathbf{v}_a \|} < \theta_{max}.$ $\qquad\qquad$ (4.2)

The visibility constraint defines "what can be seen" from a specific viewpoint and the viewing angle constraint defines "where the sensor can be placed to look at the point". They are illustrated in Fig. 4.1a and b respectively.

The sensor's acceptable pose space is a cone which has the maximum angle $V_p = \gamma(2\,\theta_{max})$. This space is called the surface point's **viewpoint space**, which is a set of 3D locations where the used sensor can be placed to take its image. Given an object to be observed, it also has a viewpoint space around the object.

4.2.3 Field of View

A common CCD camera has a field-of-view (FOV) limited by the size of the sensor area and the focal length of the lens (Abrams 1999). A surface point beyond the sensor's FOV will be projected outside the sensor area and will not be detectable. The locus, which satisfies the FOV constraint for a set of surface features, is given by:

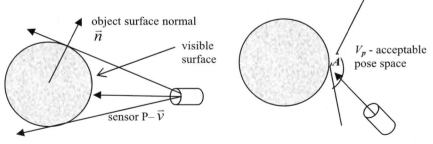

(a) What can be seen from P $\qquad$ (b) Where the sensor can be placed to look at A

Fig. 4.1. Visibility and viewing angle

$$G3: \quad \mathbf{v} \cdot \mathbf{v}_a - \| \mathbf{v} \| \cdot \| \mathbf{v}_a \| \cos(\alpha/2) \geq 0, \tag{4.3}$$

where α is the field-of-angle of the sensor (Fig. 4.2).

Since sensor planning systems generally consider the image plane to be symmetrical about the optical axis for the purposes of FOV, α is computed based on the length of the smaller side of the sensor area, I_{min}. Therefore, $\alpha = 2 \arctan(I_{min}/2d)$. For a structured light vision sensor, both the camera and the projector's FOVs need be considered.

4.2.4 Resolution

Pixel resolution can be used to determine the minimum scene feature size resolvable by the vision system. That is, to ensure that every feature is resolvable,

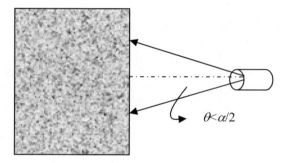

Fig. 4.2. Field of view

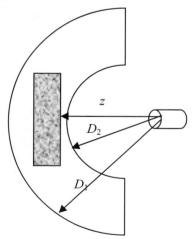

Fig. 4.3. The depth of field

one must ensure that there are at least two pixels on the image plane (Tarabanis 1995, Canceroni and Kutulakos 1999, Abrams 1999). In machine vision tasks, it is required that a particular unit feature size on an object appear as the minimum number of image elements on a sensor. This feature resolution constraint can be satisfied by properly selecting the image sensor, as well as by carefully planning its placement and settings. The objective of sensor planning for the resolution constraint is to determine the sensor parameters that achieve this resolution.

Simply, the spatial resolution is considered as the size of each pixel representing in the real world, and the constraint can be formulated as:

$$G4: \quad \sigma_{resol} = (\frac{z}{Nf} - \frac{1}{N})\frac{1}{\cos\theta} < \sigma_{acceptable} \quad (mm/pixel), \qquad (4.4)$$

where z is the distance between the lens and the object and θ is the angle between the object surface normal and the optical axis.

4.2.5 In Focus and Viewing Distance

A camera is perfectly focused at a specific distance (measured along the optical axis) given by $D=fd/(d-f)$. In practice, however, if a surface patch images into a blur circle of a given size c, it is considered sufficiently in focus for a given application. The system is then focused for a range of depths from D_1, the far limit of the depth of field, to D_2, the near limit. These limits are given by Abrams (1999):

$$D_{1,2} = \frac{afd}{a(d-f) \mp cf} \qquad (4.5)$$

where a is the entrance pupil diameter, f is the intrinsic focal length, and d is the focus distance.

Consider a digital image acquired by the vision sensor that has a size of $N \times N$ and the blur radius is restricted in on-pixel length or $(c=2a/N)$. The depth of field becomes:

$$D_{1,2} = \frac{fd}{(d-f) \mp 2f/N} \qquad (4.6)$$

Since these distances are measured along the optical axis, the corresponding constraints are given by:

$$D_1 \geq (r_f - r_v)v \quad \text{and} \quad (r_c - r_v)v \geq D_2,$$

where r_c is the feature point closest to r_v and r_f is the feature point farthest from r_v. (D_1-D_2) is the depth of field (Fig. 4.3).

If d is adjustable from d_{min} to d_{max} ($f < d_{min} < d_{max} < 2f$), the object can be put between z_{min} and z_{max}:

$$G5: \quad z_{min} < z < z_{max} \qquad (4.7)$$

where

$$z_{\min} = \frac{fd_{\max}}{(d_{\max} - f) + 2f/N}, \ z_{\max} = \frac{fd_{\min}}{(d_{\min} - f) - 2f/N}.$$

On the other hand, if the object is placed at the distance of z away from the sensor, the optimal focus distance is set to:

$$d_{\text{opt}} = \frac{zf}{z - f}. \tag{4.8}$$

For a structured light vision sensor or other 3D range sensor, this placement constraint can be simply expressed with an allowable distance range, similarly to (4.7).

4.2.6 Overlap

During the 3D reconstruction, the overlap constraint ensures that the resampled part of the object surface is already partially seen. Because the best performing registration algorithms (Higuchi et al. 1995, Okamoto 1998) and many algorithms for integrating range data perform best when the range data overlaps, registration and integration impose an overlap constraint on the choice of the next viewpoint. The size of the overlap area is dependent on the image-merging algorithm. Here we assume that it needs a minimal width,

$$\text{G6:} \quad w > w_{\min}. \tag{4.9}$$

If the robot's hand and eye are calibrated previously, $w_{\min}$ may be zero and the overlap constraint can be ignored.

4.2.7 Occlusion

Occlusion (Fig. 4.4) is an important scene attribute relative to the sensor planning process. The planning process relies on the construction of a visibility volume V_t for the target in which the sensor positions have an unoccluded view of the target. This can be computed by determining V_p, the visibility volume for the case where there are no occlusions, and subtracting O_i, the volume containing the set of sensor positions occluded from the target by model surface i, for each surface of the model (Reed et al. 1997):

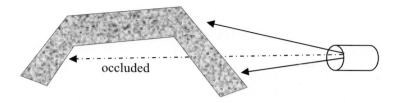

Fig. 4.4. Occlusion

$$V_t = V_p - \bigcup_{i \neq p} O_i \qquad (4.10)$$

In another way, the target A is visible by the sensor if no other entities are located between them. Here the "entity" means any geometrical element e_j, such as line, surface, or solid object. This yields

$$\text{G7:} \quad A = \begin{cases} \text{visible}: & if\,((L_{\mathrm{PA}} \cap (\bigcup_{j=1}^{n} e_j)) = \phi) \\[2mm] \text{occluded}: & if\,((L_{\mathrm{PA}} \cap (\bigcup_{j=1}^{n} e_j)) \neq \phi) \end{cases}, \qquad (4.11)$$

where $A \in \{e_j\}, j=1, 2, \ldots, n$ are total entities in the universe space; L_{PA} is the line set from the sensor's viewpoint to the target A; ϕ means an empty set of the entities intersection.

For freeform objects, the computation of occlusion constraints is very time-consuming. The next section will give a simple method to determine the occlusion for some simple entities.

4.2.8 Image Contrast

Contrast is a criterion of image quality. During the sensor planning, it may also affect the position and optical settings of the vision sensor, e.g. the diameter of the lens aperture and the illumination intensity of the projector. More details will be introduced in Chap. 10 (G8: equation 10.3).

4.2.9 Robot Environment Constraints

Kinematic reachability of sensor pose and robot-environment collision constraints should also be considered in a real robot system. Let the universal space be:

$$V = V(x, y, z, \alpha, \beta, \gamma), \ (x, y, z) \in R^3, \ -\pi \leq \alpha, \beta, \gamma \leq \pi, \qquad (4.12)$$

and the reachable space of the robot be $V_r = V_r(x, y, z, \alpha, \beta, \gamma) \subset V$. The three rotational angles are also included because some places are restricted with specified arm directions and the number of the robot's DOF may be less than six.

Definition 4.1 (collision free). Let p_1 and p_2 be two proposed sensor poses, $p_1, p_2 \in V$. If there is a spatial path for the robot eye to move from p_1 to p_2, it is said to be collision-free from p_1 to p_2, and denoted as:

$$\overrightarrow{p_1 p_2} = \overrightarrow{p_2 p_1} = 1, \qquad \text{(TRUE)}.$$

Otherwise

$$\overrightarrow{p_1 p_2} = \overrightarrow{p_2 p_1} = 0, \qquad \text{(FALSE)}. \tag{4.13}$$

Because $\overrightarrow{p_1 p_2}$ always equals to $\overrightarrow{p_2 p_1}$, both are denoted as $\overrightarrow{p_{1,2}}$ for convenience.

Definition 4.2 (shortest collision-free path). If p_1 and p_2 are both reachable by the robot eye, $p_1, p_2 \in V$, there must exist a path for the hand moving from p_1 to p_2, and $\overrightarrow{p_{1,2}}$. Denote the shortest path

$$l_{1,2} = \hbar(x, y, z, \alpha, \beta, \gamma), \qquad \hbar_{\text{start}} = p_1, \hbar_{\text{end}} = p_2, \tag{4.14}$$

Here not only the sensor position but also the sensor orientation is considered because the orientation may also affect the path planning and cause additional cost for robot operation or motion. However, some other sensor optical parameters (a, d, f) are not necessary because they do not affect the path from one viewpoint pose to another and they can be simultaneously adjusted during the time of robot motion. The length of the shortest collision-free path is called *viewpoint distance*.

Let $G = G\{P_{ci} \mid i = 1, 2, ..., n\}$ be a finite set of sensor poses related to each other for a given vision task. With any pair $i, j \in [1, n]$, if we have

$$\overrightarrow{p_i p_j} = 1, \qquad i, j \in [1, n], \tag{4.15}$$

then G is a connected graph, denoted as $\hat{C}(G) = 1$ (TRUE). That is, for any pair of sensor poses, there exists a path connecting them.

To satisfy the robot environment constraints, it is required to generate a connected graph G:

$$\text{G9:} \quad \hat{C}(G) = 1. \tag{4.16}$$

Definition 4.3 (distance of viewpoints). p_1 and p_2 are two planned viewpoints in the robot workspace. Define the viewpoint distance as:

$$w_{1,2} = \int_{p_1}^{p_2} \hbar(x, y, z, \alpha, \beta, \gamma) d\hbar \tag{4.17}$$

Details of computing the viewpoint distance are about to be presented in Chap. 6.

4.3 Common Approaches to Viewpoint Evaluation

The viewpoint evaluation criterion is very important in the strategy for determining "where to look next" (Ye 1995), which decides the selection of view candidates for the next action. In most related works, sensor system limitations are expressed as a cost function and the aim of sensor planning is to reach the goal (vision task) with minimal cost. The cost function should have a value tending to infinity associated to the direction of a pose that the sensor cannot assume and a unitary value associated to the direction of a pose that is possible for the sensor to assume (Okamoto 1998).

A common evaluation function has many adverse features to resolve, as described in Triggs (1995): (1) evaluation is relatively expensive owing to the large amount of geometric computation; (2) the evaluation function is highly nonlinear; (3) a search over all 3D sensor poses would be six-dimensional or nine-dimensional or even more; and (4) computation becomes more complex when overlap and illumination are considered.

In Banta (1996), the term "best-next-view" is defined as the next sensor pose which will acquire the greatest amount of previously unseen three-dimensional information

$$M_{p(t+1)} = \max \{ M_{p(k)} \}.$$

With this definition, however, the final trajectory may be ineffective in dealing with the final distance covered by the camera. Furthermore, such a strategy does not take into account some problems like the manipulator kinematics constraints or geometric constraints.

Some researchers (Tarabanis 1995, Gu et al. 1999) chose to formulate the probing strategy as a function minimization problem. This should define a function to be minimized which integrates the constraints imposed by the robotic system and evaluates the quality of the viewpoint. The optimization function is taken to be a weighted sum of several component criteria, each of which characterizes the quality of the solution with respect to each associated requirement separately. Thus the optimization function is written as:

$$h = \max(\alpha_1 g_1 + \alpha_2 g_2 + \alpha_3 g_3 + \alpha_4 g_4)$$

subject to $g_i{\geq}0$, i=1, 2a, 2b, 3, 4 and g_5=0, where g_1 is the resolution constraint, g_{2a} and g_{2b} are the focus constraints, g_3 is the field-of-view constraint, g_4 is the visibility constraint, and g_5 is an equality for the optimization constraint.

In Marchand (1997), the strategy of viewpoint selection takes into account three problems: quality of a new position; displacement cost; and additional constraints. More precisely, it includes: the new observed area volume $G(\phi_{t+1})$, the cost function F in order to reduce the total camera displacement $C(\phi_t, \phi_{t+1})$, and the constraints to avoid unreachable viewpoints and to avoid positions near the robot joint limits $B(\phi)$. The cost function F_{next} to be minimized is defined as a weighted sum of the different measures:

$$F(\phi_{t+1}) = A(\phi) + a_1 G(\phi_{t+1}) + a_2 C(\phi_t, \phi_{t+1}) + a_3 B(\phi) \cdot$$

Zha et al. (1997, 1998) evaluated the suitability of all potential viewpoints of the NBV by using a rating function as

$$f(\theta, \phi) = w_e f_e(\theta, \phi) + w_o f_o(\theta, \phi) + w_s f_s(\theta, \phi) \cdot$$

where θ and ϕ are two parameters on the viewpoint sphere; f_e, f_o, f_s are factor functions rating on some physical or heuristic constraints, and w_e, w_o, w_s are weighting coefficients. The viewpoint of the largest value of $f(\theta, \phi)$ will be chosen as the NBV. The definitions of f_e (extending constraints), f_o (overlapping constraints), f_s (smoothness constraints) and weighting coefficients can be found in Zha (1997, 1998).

Ye et al. (1995) argued that the total cost for applying the searching effort allocation is:

$$T[F] = \sum_{i=1}^{k} t_o(f_i) \cdot$$

where the cost $t_o(f)$ gives the total time needed to (1) manipulate the hardware to the status specified by f; (2) take a picture; (3) update the environment and register the space; (4) run the recognition algorithm. Because (1)–(3) are constant, $t_o(f)$ is only influenced by (4). Let O be the set of all the possible operations that can be applied. The effort allocation $F=\{f_1, \dots, f_k\}$ gives the ordered set of operations applied in the search. The next action is selected that maximizes the term

$$E(f) = \frac{P(f)}{\Delta_T(f)}, \qquad \Delta_T(f) = t_o(f) \cdot$$

where $P(f)$ is the probability of detecting the target.

Triggs et al. (1995) gave a method of the optimization technique to minimize their viewpoint evaluation function. They divided the search space into a set of local regions and built a probabilistic function interpolation or subjective probability distribution for the function value. These distributions can be used to choose which region to refine and where to subdivide it. The goal is to optimize the function, so a sample only "succeeds" if it improves on the best currently known function value f_{best}. If the probability density for the function value at some point is $p(f)df$, the expected gain or improvement to f_{best} from a sample placed at that point is

$$< \text{gain} >= \int_{-\infty}^{f_{best}} (f_{best} - f) p(f) df \cdot$$

However, in these previous methods, the constraints in sensor placement are expressed as a cost function with the aim to achieve the minimum cost and the evaluation of a viewpoint is usually achieved by direct computation. Such an approach is normally formulated for a particular application and is therefore difficult to be applied to general tasks. In this book, a lowest traveling cost is proposed to evaluate a sensor placement plan. It takes advantage of a minimum number of viewpoints and a shortest path through them.

4.4 Criterion of Lowest Operation Cost

As discussed in the previous section, for a feasible viewpoint, a number of the placement constraints have to be satisfied. Figure 4.5 illustrates a subset of the sensor placement constraints (G_1, G_2, G_3, G_5, G_7). Considering the 6 points (A – F) on the object surface, only point A satisfies all the 5 constraints, while all other points violate one or more of the constraints.

On the other hand, since a viewpoint can be placed randomly in the acceptable viewpoint space, a cost function is usually used to evaluate its goodness, which leads to an optimal solution of the sensing plan in a certain vision task. The term "optimal" refers to a maximal benefit for image acquisition and subsequent image evaluation for robot control steps (Kececi 1998). However, a common evaluation function has many adverse features to resolve (Triggs 1995). Previous approaches to viewpoint evaluation have been summarized in a previous section. In this book, the lowest traveling cost is used as a criterion to evaluate the sensing plan.

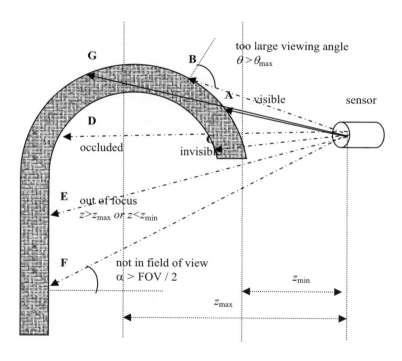

Fig. 4.5. An example of sensor placement constraints

For a task where there exists a priori model, the strategy is described as follows:

1. Generate a number of viewpoints.
2. For every feature point, test if there exist possible viewpoints in the acceptable space to observe it. If no viewpoint is possible, the feature point needs to be eliminated from the task.
3. Construct a graph corresponding to the topology of viewpoints. Try to change the viewpoints' parameters to satisfy all above-mentioned constraints.
4. Try to reduce redundant viewpoints or the order of graph.
5. Compute the lowest traveling cost to optimize robot operations, which corresponds an optimal Hamilton cycle in the graph for this problem.

Generating a large number of viewpoints will most likely satisfy all constraints for the vision task, but it will also increase the operation cost. To achieve an optimal solution, it is necessary to eliminate all possible redundant viewpoints. Figure 4.6 illustrates that the 2nd viewpoint is redundant because it does not increase any information on the object model, i.e. $M_{p2} - M_{p2} \cap (M_{p1} \cup M_{p3}) = 0$.

Definition 4.4 (sensor placement graph). The plan of viewpoints is mapped to a graph $G = (V(G), E(G), \psi_G, w_E)$ with weight w on every edge E, where vertices V_i represent viewpoints, edge E_{ij} represent the shortest collision-free path between viewpoint V_i and V_j, and edge weight w_{ij} represents their viewpoint distance. Such a graph is called as *sensor placement graph G*.

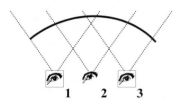

Fig. 4.6. A redundant viewpoint

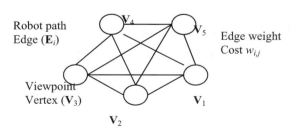

Fig. 4.7. A sensor placement graph

Figure 4.7 illustrates an example of the sensor placement graph. A practical solution of the sensor placement problem must provide the exact number of viewpoints and every viewpoint must be reachable to the robot. According Definition 4.1 and Definition 4.2, it is obvious that there must be a collision-free path between each pair of viewpoints, i.e. $\overrightarrow{p_{i,j}} = 1$, $\forall i, j \in [1, n]$. A sensor placement graph G has the following characteristics:

− G is a simple undirected graph, i.e. there are no loops and no paralleled edges;
− G is a connected graph, i.e. there is at least a path from V_i to V_j;
− G is a complete weighted graph, i.e. every pair of vertices V_i to V_j is directly connected by a weighed edge;
− G is a finite nontrivial graph, i.e. $1 \le o(G), \partial(G) < \infty$;

− The order and size of G are $o(G) = n$, $\partial(G) = \frac{1}{2}n(n-1)$, respectively.

Later we will show that the shortest path for taking views is a Hamilton cycle which is a sequence of vertices: $C = (x_1, x_2,\ldots, x_n, x_1)$ where $xi \ne xj$, $xi \in V(G)$, $i \in [1, n]$. The length of the path is

$$l_c = w(x_n, x_1) + \sum_{i=1}^{n-1} w(x_i, x_{i+1}), xi \in V(G) \cdot$$

If we consider the time required for a plan of sensor placement, it may include:
1. $n*t1$ – time needed to acquire the current view and transfer it to a depth image (3D local model), including image digitalization, coordinate system transformation, image preprocessing, corresponding (applied to stereo vision), 3D surface reconstruction, and so on.
2. $n*t2$ – time for fusion and registration, i.e. for merging the local model with the previous partial model.
3. $t3$ – time needed to perform the strategy of viewpoints planning. The viewpoints and optical settings of the sensor are determined and a path is generated for the robot. Here we assume that the hand and eye of the robot have already been calibrated previously and there exist confirmed matrixes for coordinate system transformation.
4. $t4$ – time needed for the robot to perform the task of moving from one viewpoint to another.

Here $t3$ is subject to the following constraints: ("≥" means the constraint condition is satisfied.)

$\qquad g_1 \ge 0$ (resolution constraint)
AND $\quad g_2 \ge 0$ (in-focus for stereo or viewing distance for range sensor)
AND $\quad g_3 \ge 0$ (field of view)
AND $\quad g_4 \ge 0$ (visibility)
AND $\quad g_5 \ge 0$ (viewing angle)
AND $\quad g_6 \ge 0$ (overlap for 3D reconstruction task)

AND $g_7 \geq 0$ (occlusion)
AND $g_8 \geq 0$ (image contrast)
AND $g_9 \geq 0$ (kinematic reachability of sensor pose).

If n viewpoints/actions of image acquisition are needed to finish the whole task, the total needed time is

$$\text{TaskTime} = (t1 + t2) * n + t3 + t4. \tag{4.18}$$

Since the planning strategy can run offline, t_3 can be omitted from the above equation because it is not considered during the robot operations. Assuming that t_1 and t_2 are constants and t_4 is proportional to the path length of robot the end-effector, we have

$$T_{cost} = (T1 + T2)n + l_c \kappa \tag{4.19}$$

where l_c is the path length through all the viewpoints.

It is obvious that reducing the number of viewpoints will improve the vision perception behavior. Therefore our first objective is to take the lowest traveling cost T_{cost} through the planned viewpoints. In fact, if both the object model and the robot environment are specified, the length of the shortest path of taking views will not vary very much and then the traveling cost is just proportional to the number of viewpoints. Hence the objective becomes minimizing the number of viewpoints and an optimal solution of sensor placement contains the least viewpoints and the corresponding graph has the lowest order, i.e. $o(G_{optimal}) = n_{min}$ and $l_{optimal} = l_{cmin}$.

4.5 Summary

In this chapter, the frequently used sensor constraints were formulated to limit a viewpoint to be in the acceptable space feasible for robot execution. An evaluation criterion is used to achieve a good sensor placement plan. For model-based robot vision, the sensing plan is a graph spatially distributed around the object. For nonmodel-based object modeling tasks, the planning strategy determines where to look next.

The method for model-based sensor placement achieves both the optimal sensor placements and the shortest path through these viewpoints. The plan for such sensor placements is evaluated with three factors: low order, high precision, and satisfying all constraints.

Chapter 5
Model-Based Sensor Planning

This chapter presents a method for automatic sensor placement for model-based robot vision. Since the sensor is moved from one pose to another around the object to observe all features of interest, this allows multiple 3D images to be taken from different vantage viewpoints. The task involves determination of the optimal sensor placements and a shortest path through these viewpoints. During the sensor planning, object features are resampled as individual points attached with surface normals. The optimal sensor placement graph is achieved by a genetic algorithm in which a min-max criterion is used for the evaluation. One shortest path is determined by Christofides algorithm. A Viewpoint Planner is developed to generate the sensor placement plan. It includes many functions, such as 3D animation of the object geometry, sensor specification, initialization of the number of viewpoints and their distribution, viewpoint evolution, shortest path computation, scene simulation of a specific viewpoint, parameter amendment. Experiments are also carried out on a real robot vision system to demonstrate the effectiveness of the proposed method.

5.1 Overview of the Method

The following sections present a method of automatic sensor placement for planning model-based vision tasks, typically for industrial inspection, with both optimal viewpoint distribution and sensing sequence. The procedures for generating a sensing plan include

1. Input the object's geometric information from a model database;
2. Give the specifications of the vision tasks and sensor configurations;
3. Generate a sensor placement graph with the fewest viewpoints;
4. Search for a shortest path for robot execution; and
5. Output the sensing plan.

Therefore, the problem of sensor placement for model-based vision tasks becomes the search for an optimal placement graph and a shortest path for achieving the sensing operations. In this chapter, the geometric information of the object is loaded from a 3D CAD data file. A strategy is developed to automatically determine a group of viewpoints for a specified vision-sensor with several placement parameters such as position, orientation, and optical settings. Each viewpoint should satisfy multiple constraints due to the physical and optical properties of the sensor, scene occlusion, and robot reachability in the environment etc.

As stated previously, reducing the number of viewpoints will improve the vision perception performance. In this chapter, the goal is to achieve the lowest traveling cost T_{cost} through the planned viewpoints, but not by a combined function as in traditional approaches. It will be achieved by (1) minimizing the number of viewpoints subject to task completion, (2) optimizing the viewpoint distribution, and (3) finding one shortest traveling path.

5.2 Sensor Placement Graph

5.2.1 HGA Representation

Here the hierarchical GA is used to determine the optimal topology in the sensor placements which will contain the minimum number of viewpoints with the highest accuracy while satisfying all the constraints. The hierarchical chromosome can be regarded as the DNA that consists of the parametric genes and control genes. In this work, parametric genes (V_i) mean the sensor poses and optical settings, and control genes (c_i) mean the topology of viewpoints. To show the activation of the control gene, an integer "1" is assigned to each control gene being enabled whereas "0" indicates a state of turning off. When "1" is signaled, the associated parameter genes associated with that particular active control gene are activated in the lower-level structure.

For the sensor placement problem, a chromosome in GA represents a group of viewpoints with a specific topology. Figure 5.1 illustrates the structure of the hierarchical chromosome corresponding to the plan of viewpoints. Here V_i represents a variable viewpoint which is a vector of the sensor parameters. E.g., for the structured light system, $V_i = (x, y, z, \alpha, \beta, \gamma, b, \psi, a)$, and for a stereo sensor, $V_i = (x, y, z, \alpha, \beta, \gamma, a, f, d)$. The corresponding $c_i = \{0, 1\}$ represents a control gene which is a binary variable.

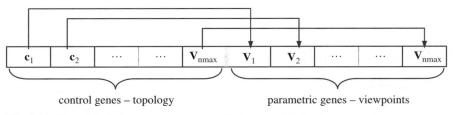

Fig. 5.1. Hierarchical chromosome structure for sensor placement

5.2.2 Min-Max Objective and Fitness Evaluation

A plan of sensor placements is evaluated by a min-max criterion, which includes three objectives and a fitness evaluation formula. The order of a graph G is equivalent to the number of occurrences of "1" in the control level genes. To plan a group of viewpoints with minimum order in the sensor placement graph, the first objective is given as:

$$\text{Objective 1:} \quad \text{minimize } o(G) = \sum_{i=1}^{n\,\max} c_i \,.\tag{5.1}$$

Assume that the accuracy of vision inspection is proportional to the surface resolution of the vision sensor and consider m features to be acquired. The second objective is to improve the average accuracy via

$$\text{Objective 2:} \quad \text{maximize } \eta(F) = \frac{1}{m}\sum_{j=1}^{m}\frac{w_{j,\,\text{image}}}{l_j}\tag{5.2}$$

where w_{image} is the size of a feature on the sensor image and l_j is its actual length.

On the other hand, an admissible viewpoint is subject to some constraints in the sensor placement space, e.g. resolution, in-focus, field of view, visibility, viewing angle, overlap, occlusion, contrast, and reachability. A penalty scheme is set up to handle these constraints such that invalid chromosomes become low performers in the population. The constrained problem is then transformed to an unconstrained condition by associating the penalty with all the constraint violations. A vector of penalty coefficients combines the nine constraints (presented in Chap. 4):

$$K = (\delta_1, \delta_2, \delta_3, \delta_4, \delta_5, \delta_6, \delta_7, \delta_8, \delta_9) \,.\tag{5.3}$$

where δ_i is the constant weight representing the importance of that constraint. If the constraint does not need to be satisfied (e.g. G_6-overlap), the weight is set to zero.

Define a binary function

$$\varphi_i = \begin{cases} 0, & \text{if the constraint is satisfied} \\ 1, & \text{if the constraint is violated} \end{cases}\tag{5.4}$$

and construct another vector of constraints:

$$Q(l, V) = (\varphi_1, \varphi_2, \varphi_3, \varphi_4, \varphi_5, \varphi_6, \varphi_7, \varphi_8, \varphi_9) \,,\tag{5.5}$$

where l is an object feature and V is a viewpoint.

Therefore the third objective is to minimize the total penalties for the constraints:

$$\textbf{Objective 3:} \quad \textbf{minimize } \text{penalty} = K \cdot Q^T \,.\tag{5.6}$$

If there are m features and n viewpoints, the average penalty for a viewpoint topology is:

$$\text{penalty} = \frac{1}{m} \sum_{i=1}^{m} \min[K \cdot Q(l_i, V_j)^T \mid j = 1, 2, ..., n] \cdot \tag{5.7}$$

Finally, the fitness function is derived by combining the penalty scheme with the two objective functions:

$$\textbf{Fitness:}\quad \textbf{f(G)} = (an_{max} + b\ell_{max} + \mid K \mid)$$

$$- ao(G) - \frac{b}{\eta(F)} - K \cdot Q^T, \tag{5.8}$$

where $\mid K \mid = \sum_{i=1}^{m} \delta_i$, $(an_{max} + b\ell_{max} + \mid K \mid)$ is the maximum possible value that produces positive fitness, ℓ_{max} is the maximum possible resolution, and a and b are two adjustable scaling factors.

5.2.3 Evolutionary Computing

According to the characteristics of the sensor placement problem, the following genetic parameters and operations are adopted for evolutionary computing:

- Chromosome length:
- $2n$, where n is the maximum number of viewpoints and determined according to the object size and sensor configurations,
- Crossover method:
- control level genes: one-point crossover if $n < 10$, two-point crossover if $n >= 10$; probability of crossover $p_c = 0.25$;
- parametric level genes: Heuristic crossover with ratio=0.8. Here the parameters of the sensor pose and optical settings are real numbers.
- Mutation method:
- control level genes: bit-flip mutation; probability of mutation $p_m = 0.01$.
- parametric level genes: $g = g + \phi(\mu, \sigma)$ where ϕ is the Gaussian distribution function, μ and σ are the mean and variance, respectively.
- Selection method: Roulette-Wheel selection method;
- Replacement: Steady State without duplicates;
- Population size: 30–100, based on the length of chromosome;
- Initial population: randomly generated on a sphere around the object.

5.3 The Shortest Path

5.3.1 The Viewpoint Distance

For a sensor placement graph, there may exist more than one path with the shortest (or approximately shortest) length through all the viewpoints. To determine a shortest path, we firstly need to compute the distance, $w(V_i, V_j)$, between each pair of viewpoints (V_i and V_j). The concept of viewpoint distance is defined in Chap. 4. With different types of robots and different control modes, the distance should be computed in different ways accordingly.

5.3.1.1 Tool Level Control

To achieve a robot-independent representation of the location of the robot tool or hand, the control program often defines the locations in terms of a Cartesian reference frame fixed to the base of the robot or workspace. If the robot moves at a constant speed, the execution time is proportional to the 3D position difference (Euclidean distance) or 3-axis orientation difference, i.e.

$$w(V_i,V_j) = \mathbf{max}(\|^{xyz}V_i - ^{xyz}V_j\| / \boldsymbol{\mu}, \|^{\alpha\beta\gamma}V_i - ^{\alpha\beta\gamma}V_j\| / \mathbf{v}), \tag{5.9}$$

where μ and v represent the translational speed and rotational speed respectively, ^{xyz}V and $^{\alpha\beta\gamma}V$ are vectors for the three position and orientation components of V, respectively.

5.3.1.2 Asynchronous Joint Control

If the robot is controlled asynchronously in a joint space to change its pose, the distance is computed by

$$w(V_i,V_j) = \sum_{t}^{n_{dof}} (\|^t V_i - ^t V_j\| / \mu_t), \tag{5.10}$$

where n_{dof} is the robot's DOF number, μ_t is the execution speed of joint t, and $^t V$ is the joint position at pose V.

5.3.1.3 Synchronous Joint Control

When the robot is controlled in a joint space with all joints moved simultaneously, the distance is determined by the maximum one

$$w(V_i,V_j) = \mathbf{max}\{\|^t V_i - ^t V_j\| / \mu_t\}, t \in [1, n_{dof}]. \tag{5.11}$$

If the sensor's optical settings (e.g. zoom, focus, Iris, etc.) are under motorized control, it will also take time to change the sensor configuration from one viewpoint to another. Then the viewpoint distance will be the larger of this distance (time equivalent) and the above robot pose distance.

With n viewpoints obtained by generic algorithm, a symmetrical distance matrix can be generated by computing each pair of the viewpoints: $\mathbf{W} = \{w_{ij}\}$, which will be used to determine the shortest path.

5.3.2 Determination of a Shortest Path

Assume that the robot should resume its initial state after completing a vision task (since it needs be ready for inspection of the next workpiece). Given a specified graph, now another fundamental task is to find an optimal closed chain that is the shortest (or approximately shortest) one of all the possible chains. Obviously a sensor placement graph satisfies the triangle inequality, i.e.

$$w(V_i,V_j) \le w(V_i,V_k) + w(V_k,V_j), \forall V_k \in V(G) \setminus \{V_i,V_j\}, \tag{5.12}$$

where the "=" holds if the position of V_k is on the path l_{ij} and the orientation of V_k is the middle angle between Ω_i and Ω_j.

Because a sensor placement graph G is finite, connected, and complete, the optimal closed chain is the optimal Hamilton cycle. Furthermore, a complete graph must contain Hamilton cycles, i.e. there exist cycles which contain all the vertices once. In graph theory, it has been proved that if G is complete and satisfies the triangle inequality, the optimal chain C'' in a connected and weighted graph G'' corresponds to an optimal cycle C in its complete and weighted graph G. That is, $C'' \leftrightarrow C$ and $w(C'') = w(C)$, where $w(X)$ is the length of chain or cycle X.

To plan a sequence of robot operations or to find an optimal Hamilton cycle, we have to decompose G_n into the union of some edge-disjoint Hamilton cycles. There are totally n vertices and $\partial(G) = \frac{1}{2}n(n-1)$ edges in the graph G_n. A Hamilton cycle C must contain n edges too. Let a Hamilton cycle be a sequence of vertices: $C = (x_1, x_2, ..., x_n)$ where $xi \ne xj, xi \in V(G), i \in [1,n]$. The problem might be solved by enumerating all possible Hamilton cycles C_i in the graph, by comparing their summed weights $w(C_i)$, and then finding out the smallest one $cost = \min[w(C_i)]$. However, there are totally $o(C) = \frac{1}{2}(n-1)!$ Hamilton cycles. When n is large, this will yield unacceptable computations, e.g. $o(C) = 6 \times 10^{16}$ when $n = 20$. This is a non-deterministic polynomial complete (NPC) problem in graph theory and must be solved by an approximation algorithm.

This chapter uses an approximation algorithm developed by Christofides. The procedure of this algorithm for finding an optimal Hamilton cycle is described as:

− Step 1. Construct the distance matrix $\mathbf{W}$ from graph (G, w).
− Step 2. Find the smallest tree $\mathbf{T}$ in $\mathbf{W}$ using *Prim algorithm*
− Step 3. Find the odd degree set V in $\mathbf{T}$ and calculate the *perfect matching* $\mathbf{M}$ of the smallest weights in $G[V]$ using *Edmonds-Johnson algorithm*.

- Step 4. Find an Euler circuit $C_0=(x1, x2, x3, ..., xn)$ in $G^*=T+M$ using *Fleury algorithm*.
- Step 5. Start from $x1$ and go along C_0, remove each multi-occurrence vertex from C_0 except for the last $x1$ and finally form a Hamilton cycle C of graph G. This gives the approximated optimal cycle.

The resulting Hamilton cycle is an approximate solution. It has been proven that the error ratio does not exceed 0.5 even in the worst case. If L_0 is the optimal solution (sum of weighs) and L is the approximate solution by Christofides algorithm, we have $1 \leq L/L_0 \leq 1.5$. In this algorithm, the total computation cost is $O(n^3)$. In contrast, using direct search method takes $O(n!)$.

5.4 Practical Considerations

5.4.1 Geometry Scripts

The object model is usually extracted from a CAD file. However, directly using these data may result in prohibitive computation for planning the sensor placements. For example, Prieto et al. (1999) chose to import the CAD model with IGES format, which contains the NURBS representation of object surfaces. This data format must be converted to 3D voxels so that they can search for a viewpoint set. Even with a very simple object, a large number of 3D voxels will result, making the computation too costly.

This book defines a format of "Geometry Scripts" (GS) in which the whole object is constructed with some geometric primitives, such as surfaces, solid boxes, cylinders, spheres, etc. These primitives are used to build higher-level geometries by CSG (Constructive Solid Geometry) operations, such as AND ("*"), OR ("+"), and SUB ("–"). Using geometry scripts has two advantages: (1) it is intuitive and easy to understand. Users may directly write (instead of import from a CAD file) the scripts to describe what needs to be inspected. (2) The more important advantage is that it is very convenient for computing the sensor placements. A point can be checked as to whether it satisfies the placement constraints within a short period of time.

5.4.2 Inspection Features

Any geometry elements/entities, such as cylinders, freeform surfaces, curves, and individual points, can be specified as the features which need to be inspected in the vision task. However, for computation simplicity, all these features will firstly be converted into individual points. In this research, this is accomplished by a resampling method and usually the normal of each point can also be determined automatically. Figure 5.2 illustrates an object sampled with about one thousand

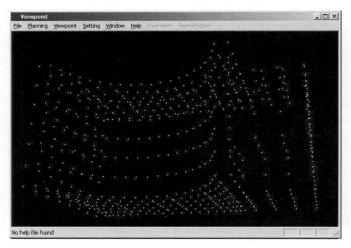

Fig. 5.2. Sampling of inspection features

points. The sampling rate was determined automatically according to the sensor configuration and surface size.

5.4.3 Sensor Structure

For a 3D sensor with two components, e.g. a structured light system with a camera and a projector or a stereo vision sensor with two cameras, the sensor placement constraints should be satisfied by each individual component. To reduce the computation complexity, we may rebuild an equivalent 3D sensor.

The projector can also be considered as a camera. Figure 5.3 illustrates a 3D sensor with a small angle between the two components. It is equivalent to a single camera placed at O_{eh} and with an effective field of angle $\text{Fov}_{eh} = \text{Fov} + \theta_r$. More conveniently, it is equivalent to be placed at O_e with FOV:

$$\text{Fov}_e = 2\tan^{-1}[(1 - \frac{h}{d})\tan(\text{Fov} + \theta_r)]$$
$$= 2\tan^{-1}[\tan(\text{Fov} + \theta_r) - \frac{b}{2d} - \frac{b}{2d\cos(\text{Fov} + \theta_r)}]$$

(5.13)

where θ_r is the relative angle between the two components, b is the baseline length, and d is the viewing distance. This formulation assumes that the two components have the same field of view. If they are different, (5.13) should be revised accordingly.

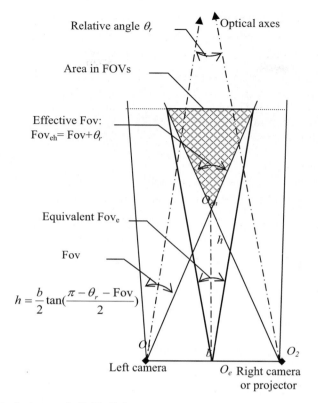

Fig. 5.3. Equivalent sensor's field of view

5.4.4 Constraint Satisfaction

In principle every constraint has to be satisfied for a viewpoint in a plan. If the viewpoint violates any of the constraints, it should be rejected (during the evolution it may also be kept according to the overall fitness). The calculation order of the constraints is important for improving the computation efficiency. This research uses the following computation order: visibility, field of view, viewing angle, in-focus, occlusion, and others. When one of them is violated, the other constraints will not be checked. Of all sensor placement constraints, the visibility has the lowest computation cost, just involving the calculation of $N_i V_i < 0$. It is thus checked as the first constraint.

The occlusion constraint is the most complex one and takes most of the computation time. To test if a point is occluded by other geometry elements, we need to check if the line segment between the point and the sensor intersects with these elements. For some regular geometry elements, such as ball, circle, square, box, etc., this can be tested in a simpler way. Taking the example of squares, this can be achieved via the following steps:

1) Assume:
   object point $\mathbf{Q} = (x_q, y_q, z_q)$, with normal $\mathbf{N} = (l_q, m_q, n_q)$;
   sensor position $\mathbf{S} = (x_s, y_s, z_s)$, orientation $\mathbf{T} = (l_s, m_s, n_s)$;
   square vertex $\mathbf{P}_i$, $i=1, 2, 3, 4$.
2) Translate the coordinate system to origin $\mathbf{Q}$:
   $\mathbf{Q}' = 0$
   $\mathbf{S}' = \mathbf{S} - \mathbf{Q}$
   $\mathbf{P}_i' = \mathbf{P}_i - \mathbf{Q}$ ($i=1, 4$)
3) Transform the z-axis to point it along $\mathbf{QS}$:
   $\mathbf{V}_{QS} = \mathbf{S}' / \| \mathbf{S}' \| = (a, b, c)$.
   $\mathbf{P}_i'' = \mathbf{R}\mathbf{P}_i'$,
where

$$\mathbf{R} = \begin{bmatrix} ac/\sqrt{1-c^2} & bc/\sqrt{1-c^2} & -\sqrt{1-c^2} \\ -b/\sqrt{1-c^2} & a/\sqrt{1-c^2} & 0 \\ a & b & c \end{bmatrix}. \tag{5.14}$$

4) Check the occlusion:

$$b_{occ} = (\rho_{12}\, \rho_{34} > 0) \text{ AND } (\rho_{23}\, \rho_{41} > 0), \tag{5.15}$$

where $\rho_{ij} = \mathbf{P}_i''(x)\, \mathbf{P}_j''(y) - \mathbf{P}_i''(y)\, \mathbf{P}_j''(x)$.

For freeform geometries, the occlusion has to be determined by "object projection" with "depth test" or "bounding volume test", which will be computationally expensive.

5.4.5 Viewpoint Initialization

A good initial population will improve the efficiency of the evolutionary computing. In this research, the initial guess of the maximum number of viewpoints and their space distribution is made via the following two steps.

(1) Estimation of the viewpoint number

We first compute the object's geometric center and find a least sphere to surround it. The maximum number of viewpoints is estimated by:

$$N = \frac{2S_{object}}{S_{view}} = \frac{2\pi^2 R^4}{d^2 \tan^2(\dfrac{Fov}{2})} \tag{5.16}$$

where S_{obj} is the object surface area, S_{view} is a single view size, R is the sphere radius, d is the average viewing distance.

(2) Uniform distribution on a sphere

Since a uniform distribution of an arbitrary number of viewpoints on a sphere cannot be described by a general mathematical formula, it needs to be handled by a special method. Here we adopt an artificial physics method to solve this problem. Take the viewpoints as particles with the same electric charge and randomly sprinkle them on the sphere. Each particle will repel every other particle. The system will lead to a stable (minimum energy) configuration where each particle is equidistant from all the others and each particle is maximally separated from its closest neighbors by the electric repulsive forces.

The sensor orientation is set to look inward to the sphere center, which is described by two parameters, θ and φ based on the sphere coordinate system. They can be expressed as a unit direction vector:

$$\vec{P}_d = (x\vec{\mathbf{i}}, y\vec{\mathbf{j}}, z\vec{\mathbf{k}}), \quad \text{where} \begin{cases} x = \cos\varphi\cos\theta \\ y = \cos\varphi\sin\theta \\ z = \sin\varphi \end{cases} \cdot \tag{5.17}$$

Figure 5.4 illustrates an example of the initial distribution of viewpoints on a sphere. The maximum number is determined according to the object and sensor configuration. The distribution is uniform (with similar minimum distance) on the sphere, but the positions are still random.

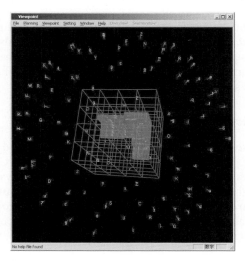

Fig. 5.4. Initial viewpoint distribution

5.5 Implementation

5.5.1 The Viewpoint Planner

In this work, a 3D animation system was developed, called *Viewpoint Planner*, which includes the following functions:

- 3D geometry input: using specially defined scripts with CSG logic operations;
- Selective display of the object, lighting effect, 3D grids, sampled inspection features, viewpoint distribution, etc.;
- Illustration of the shortest path through the viewpoints;
- Simulation of an arbitrary viewpoint to look through;
- Acquisition of the 3D map of the current view (simulation of a scene depth map);
- Configuration of sensor specifications;
- Configurable scene apparent effects: with lighting, texture, or color rendering;
- Amendment of the parameters of a specific viewpoint
- Initial estimation of the maximum number of viewpoints for an object (according to the object size and sensor configuration);
- Uniform generation of the initial viewpoints on a sphere around the object;
- Evolutionary Computation of an optimal sensor placement graph;
- Determination of a shortest path through the viewpoints;
- Viewpoint sequence export to a file.

5.5.2 Examples of Planning Results

This section presents several examples of sensor placements which were carried out with the *Viewpoint Planner*. In operations, the user only needs to give the object model scripts and sensor configurations. The system will automatically generate the initial guess, perform the evolutionary computation of optimal sensor placements, and determine one shortest path.

5.5.2.1 The Structured Light Vision Sensor

Figure 5.5 illustrates a structured light vision sensor to be used in the experiments for 3D acquisition. The sensor contains a DLP projector and CCD camera placed about 165 mm away. Figure 5.6 illustrates the sensor's equivalent configuration. The projector has the FOV of about $23°$ and is larger than that of the camera (about $15°$). Since the camera's visible volume is included by the projector, the sensor's placement constraints are equivalent to the camera's constraints and the computation is simplified to treat with a single camera.

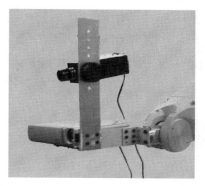

Fig. 5.5. The structured light sensor

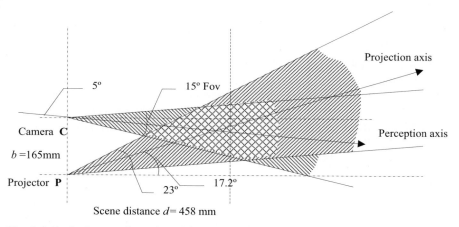

Fig. 5.6. Equivalent configuration of the structured light sensor

Three examples are given in this section. The first two examples have the same sensor configurations, but with different object models and inspection tasks. The third example has the same object as the first one, but with different sensor settings tasks. The planning results are given below.

Example One

- Target: Object 1 (shown in Fig. 5.14), 325×254×183 (mm),
- Inspection task: Full observation of all surfaces except for the bottom,
- Sensor configuration: Structured light sensor with a camera: 25 mm lens, 27.12 mm focal length, 2/3" sensor, F2.8, 15.02° Fov; and the projector: 23 mm lens, 24.22 mm focal length, 1024*768 DMD device, F3.0, 23° Fov. Equivalent sensor: 15° Fov, 424.4307–496.3449 mm field depth.

- Initialization: 60 viewpoints,
- Final optimized sensor placement plan: 28 viewpoints,
- Naive path length: 304.5132,
- The shortest path: 101.3038.

Example Two

- Target: Object 2, 300×150×180 (mm);
- Inspection task: All surfaces except for the bottom,
- Sensor configuration: Same as in Example One,
- Initialization: 42 viewpoints,
- Final optimized sensor placement plan: 25 viewpoints,
- Naive path length: 128.491744,
- The shortest path: 59.628443.

Example Three

- Target: Object 1, 325×254×183 (mm);
- Inspection task: Full observation of all surfaces except for the bottom;
- Sensor configuration: Structured light sensor with a camera: 16 mm lens, 16.58 mm focal length, 2/3" sensor, F2.8, 22.14° Fov; and the projector: 23 mm lens, 24.22 mm focal length, 1024*768 DMD device, F3.0, 23° Fov. Equivalent sensor: 22° Fov, 422.9065–494.5624 mm field depth.
- Initialization: 28 viewpoints,
- Final optimized sensor placement plan: 16 viewpoints,
- Naive path length: 207.6031,
- The shortest path: 92.3486.

The naive path length is an arbitrary path length not optimized for comparison with the shortest path found by the proposed method. These results were obtained on a PC with 750MHz CPU, 128MB RAM, under a Windows 2000 operation system. Figures 5.8–5.10 illustrate the results of viewpoint distributions and the shortest paths.

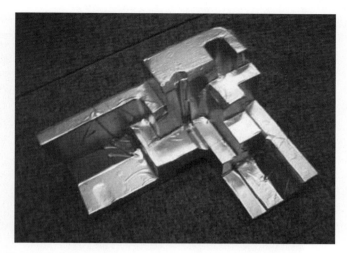

Fig. 5.7. The object to be inspected

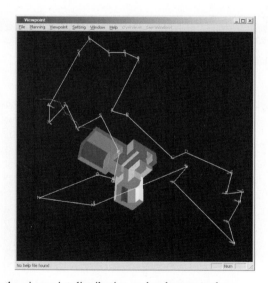

Fig. 5.8. Example 1: the viewpoint distribution and a shortest path

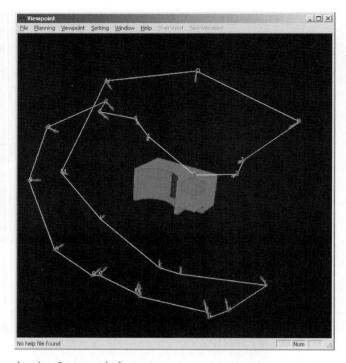

Fig. 5.9. View planning for example 2

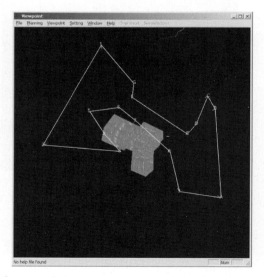

Fig. 5.10. Planning for example 3

The computation time for finding the optimal view plan in the three examples was between 1 and 3 hours, whereas the time taken for finding the shortest path was about 2 seconds. The above computation time is for reference only since the software was running in a debug mode. The actual speed should be higher. However, as the plans are generated off-line, this computation cost is not important here.

5.5.3 Viewpoint Observation

With the *Viewpoint Planner*, after evolutionary computing the user may check what can be seen from an individual viewpoint. Figure 5.11 shows the simulated scene observed from viewpoint Nos. 26 and 28 (in Example One), with the specified sensor configurations (especially the field-of-angle). Every viewpoint can be observed and the corresponding 3D depth map can also be generated for comparison with real situations.

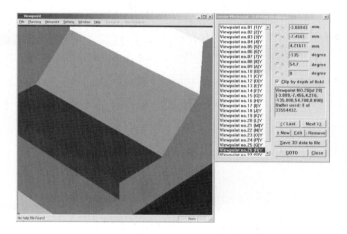

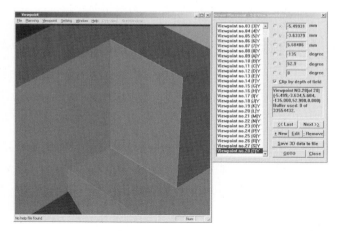

Fig. 5.11. Scenes from viewpoint Nos. 26 and 28 in the example 1

5.5.4 Experiments with a Real System

Experiments were also conducted on a real robot to verify the planning results. The system setup includes a 6DOF robot (STAUBLI RX-90B) with ±0.02 mm repeatability, a high-speed vision system, the object (illustrated in Fig. 5.7), a light source, and a sensor. We assume that the sensor placement graph has been generated at the offline stage by the *Viewpoint Planner*. In this stage, the robot is controlled to move to the specified viewpoints and the scene images are captured by the vision sensor.

Here we give the results of two typical viewpoints (Nos. 3 and 8 in Fig. 5.10) in Example 3. Figure 5.12 illustrates the view contents at viewpoint No. 3 in the Example 3. The left figure illustrates the expected scene generated by the *Viewpoint Planner*. With the structured light sensor, the 3D depth map of the view can be obtained directly. The right figure illustrates the 3D data obtained by the structured light sensor. Figure 5.13 illustrates that of viewpoint No. 8.

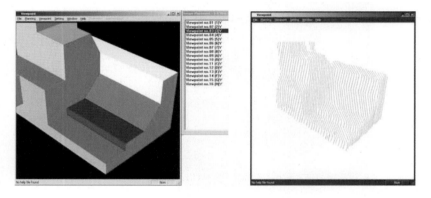

Fig. 5.12. Scene at viewpoint No. 3 in example 3 (with structured light sensor)

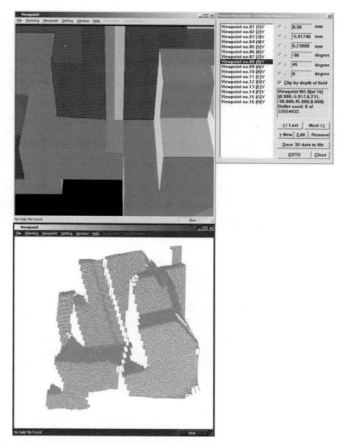

Fig. 5.13. Scene at viewpoint No. 8 in example 3 (with structured light sensor)

The experiments show that the results in real implementation match the simulation results well. Hence the sensor placement plan generated by the *Viewpoint Planner* and performed by the robot demonstrated satisfactory performance in real operations. However, the planning results may not represent the best plan. Here the results are considered acceptable or satisfactory if the viewpoints in the plan do not violate any sensor placement constraints and the redundant viewpoint number is sufficiently small. To achieve a more optimal plan in the result, longer time would be needed for the evolutionary computation.

5.6 Summary

Using conventional methods, it is difficult to achieve an optimal sensor placement graph because of the complex, large-scale, highly nonlinear characteristics of the problem. As a numerical optimizer, HGA generates the solutions that are not mathematically oriented, but possesses an intrinsic flexibility and the freedom for choosing desirable optima according to task specifications. The search for the shortest path through a number of viewpoints is an NP-Complete problem and an approximation algorithm has to be used for determining the viewing sequence. The Christofides algorithm is an effective one that can guarantee minimum error even in the worst case.

Compared with the previous approaches, this proposed method can deal with complex objects with multiple inspection features and viewpoints, generate the minimum number of viewpoints, and lead to global optimization. It provides a stable and complete solution for model-based vision tasks, including viewpoint decision, constraint satisfaction, optimization of viewpoint distribution, planning of the robot operation sequence. All these techniques are integrated into the software, the *Viewpoint Planner*, to make it useful in practical applications. The experimental results show that the real situations match the planned results well.

Therefore, the method can solve the problem of model-based sensor placement in which multiple features need to be inspected by multiple viewpoints, whilst previous researches only focused on the determination of a good viewpoint to observe a few features. However, further work can be carried out on this problem to improve the solution procedure. The GA used for determining the best viewpoint distribution usually takes a long time for searching the optimal solution. The speed is even much slower if the object is rather complex or we consider more constraints. It also cannot be guaranteed that it achieves the optimal solution. We are seeking for other optimization methods to determine the sensor placement graph for comparison. On the other hand, the Christofides algorithm can guarantee to obtain a short path within a certain error, but currently there is considerable progress in solving the shortest path problem in graph theory. Some new algorithms may be considered as alternatives to Christofides algorithm.

Chapter 6
Planning for Freeform Surface Measurement

In this chapter, we present a sensing strategy for determining the probing points for achieving efficient measurement and reconstruction of freeform surfaces. B-spline is adopted for modeling the freeform surface. In the framework of Bayesian statistics, we develop a model selection strategy to obtain an optimal model structure for the freeform surface. Based on the selected model structure, a set of probing points are then determined where measurements are to be taken. In order to obtain reliable parameter estimation for the B-spline model, we analyze the uncertainty of the model and use the statistical analysis of the Fisher information matrix to optimize the locations of the probing points needed in the measurements. Using a "data cloud" of a surface acquired by a 3D vision system, we implemented the proposed method for reconstructing freeform surfaces. The experimental results show that the method is effective and promises useful applications in multi-sensor measurements including vision-guided CMM for reverse engineering.

6.1 The Problem

Reconstructing the freeform surface from a set of discrete measurement data points is a problem important to many areas including reverse engineering, metrology, inspection by machine vision, computer-aided design (Song and Kim 1997, Thompson and Owen 1999, Wolovich et al. 2002, Weir et al. 2000). The first task in the reconstruction of a freeform surface is to obtain the measurement data. Among the various sensing techniques available, mechanical contact probes such as CMM (Coordinate Measuring Machine)'s touch probe, and 3D topography measuring systems using structured light or fringe illumination are widely used in practical applications. CMM with touch-triggered probes can provide high measurement accuracy at sub-micron level. However, the measurement speed is much lower than that of a 3D vision system. A vision system can acquire thousands of data points over a large spatial range in a snapshot (Li and Chen 2003). However, the achievable resolution is relatively low, at around 100–200 μm. Therefore, in practical applications, using one of the techniques means that the user has to suffer from its limitations, e.g. the low speed with CMM.

A way to overcome the limitations of individual sensing techniques lies in integrating multiple sensors in the measurement as conceptualized in Fig. 6.1. Research efforts have been made to achieve this. For example, Nashman et al.

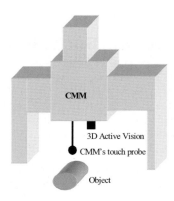

Fig. 6.1. Multiple-sensor coordinate measuring system

(1996) integrated vision in a touch-probe system, where a video camera with a laser triangulation probe and a 3D touch probe were used in a CMM. They presented a cooperative interaction method for the vision and touch-probe system that provided sensory feedback to the CMM for dimension inspection tasks. Chen and Lin (1997) presented a vision-aided reverse engineering approach (VAREA) to reconstruct free-form surface models from physical models, with a CMM equipped with a touch-triggered probe and a vision system. The VAREA integrated computer vision, surface data digitization and surface modeling into a single process. The initial vision-driven surface triangulation process (IVSTP) generated a triangular patch by using stereo image detection and a constrained Delaunay triangulation method. The adaptive model-based digitization process then refined the surface reconstruction using measurements from the CMM's touch probe. Since the vision system in VAREA used a 3D stereo algorithm to detect 3D surface boundaries, only 3D surface boundaries were reconstructed and geometrical information about the freeform surface could not be retrieved. Recently, Shen et al. (2000, 2001) presented a multiple-sensor coordinate measuring system for automated part localization and rapid surface digitization. The multiple-sensor system consists of a high-precision CMM equipped with a touch probe and a 3D active vision system. Their research focused on setting up a multiple-sensor system and processing the geometrical information from the vision system. In these systems, the CMM's touch probe plays the role of accurately digitizing a surface, especially when high-precision is desired. The question of how to determine the set of measurement data, including the needed number of the measurement data points and their locations, for accurate reconstruction of freeform surfaces, remains untouched.

Using a CMM for 3D measurements, only a finite number of discrete measurement data can be taken for a surface. From the statistical viewpoint, each measurement data point contains a certain amount of geometrical information about the surface, and the quantity of information contained in the set of measurement data points depends on the number and locations of the measurement points.

Considering the lengthy time needed in using a CMM to take a large number of measurement data points, we should select the locations of the data points to achieve an optimal measurement and reconstruction. Unfortunately, the current practice in using CMM mostly adopts random data point setting on a surface. In such a case, each data point has an equal probability of being picked for the measurement. For example, Woo et al. (1995) presented a sampling strategy based on Hammersley sequence to determine the number of discrete sample points and their locations on a machined surface (Woo and Liang 1993). Lee et al. (1997) proposed a feature-based method, which integrates Hammersley sequence and a stratified sampling method, to derive the sampling strategy for various surfaces such as circular, conic, cylindrical, rectangular and spherical surfaces.

Unlike objects composed of simple geometric primitives, such as planes, lines, spheres and cylinders, freeform surfaces have no obvious features. Therefore, they are more difficult to define and model mathematically than simple geometric objects. In most cases, freeform surfaces are represented by the parametric equations such as Coons patches (Farin 2002), B-splines, or NURBS (non-uniform rational B-splines). A fundamental question then arises: can we find the parametric model to represent an unknown freeform surface and then select a minimal set of discrete measurement points to obtain these parameters, while controlling the uncertainty of model parameters within a specified tolerance. Here, the uncertainty describes the tolerance range within which the unknown true surface lies with some confidence levels. The lower the uncertainty of the model, the better the reconstructed surface is. In this chapter, we propose a method that allows for optimal measurements and reconstruction of freeform surfaces. Two issues need to be addressed here. The first is how to select the model structure using a cloud of low-precision data acquired by a 3D vision sensor. We use B-splines to represent a freeform surface and present a Modified BIC (Bayesian Information Criterion) approach for selecting an optimal model structure for surface representation. The second is how to determine the locations of a set of measurement data points for high-precision measurements e.g. by CMM's touch probe. In our work, we analyze the uncertainty of the B-spline model and use the statistical analysis of the Fisher information matrix (Wang 1999) which measures the uncertainty of the parameters of the model, to optimize the locations of the measurement data points to minimize the uncertainty of the model.

The rest of this chapter is organized as follows. Section 6.2 describes the B-splines approximation and model selection for the 3D reconstruction of freeform surface. In Sect. 6.3, the uncertainty of the B-spline surface is analyzed. Section 6.4 presents the optimization of the locations of measurement data points. Section 6.5 gives some experimental results in reconstructing the freeform surfaces of some real objects. Finally, conclusions of the work are given in Sect. 6.6.

6.2 B-Spline Model Representation

6.2.1 B-Spline Representation

A B-spline surface is defined by the following equation

$$s(u,v) = \sum_{i=0}^{n_u-1} \sum_{j=0}^{n_v-1} B_{i,p}(u) \cdot B_{j,q}(v) \cdot \phi_{ij}. \tag{6.1}$$

where n_u and n_v are the number of control points in u and v directions; ϕ_{ij} ($i = 0,1,\ldots,n_u-1$, $j = 0,1,\ldots,n_v-1$) are the n ($n = n_u \times n_v$) control points; $B_{i,p}(u)$ and $B_{j,q}(v)$ are the normalized B-splines of degree p and q for the u and v directions respectively which are defined over the knot vectors $\mathbf{u} = [u_0, u_1, \ldots, u_{n_u+p}]$ and $\mathbf{v} = [v_0, v_1, \ldots, v_{n_v+q}]$.

Assume that (x_k, y_k, z_k) are the coordinates of a measurement point $\mathbf{r}_k$ on the surface, with location parameters (u_k, v_k). Let us further assume that the degrees of p and q and the complete knot vectors $\mathbf{u}$ and $\mathbf{v}$ for surface fitting are also determined. By introducing the measurement point $\mathbf{r}_k$ with the corresponding location parameters $[u_k, v_k]$ in (6.1), we have

$$\begin{cases} x_k = \displaystyle\sum_{i=0}^{n_u-1} \sum_{j=0}^{n_v-1} B_{i,p}(u_k) B_{j,q}(v_k) x_{ij} \\ y_k = \displaystyle\sum_{i=0}^{n_u-1} \sum_{j=0}^{n_v-1} B_{i,p}(u_k) B_{j,q}(v_k) y_{ij} \\ z_k = \displaystyle\sum_{i=0}^{n_u-1} \sum_{j=0}^{n_v-1} B_{i,p}(u_k) B_{j,q}(v_k) z_{ij} \end{cases} \tag{6.2}$$

where (x_{ij}, y_{ij}, z_{ij}) are the coordinates of the B-spline surface control points ϕ_{ij}. (6.2) can be expressed as a linear combination of the control points in the B-spline representation,

$$\begin{cases} x_k = \mathbf{B}_k \cdot \mathbf{\Phi}_x \\ y_k = \mathbf{B}_k \cdot \mathbf{\Phi}_y \\ z_k = \mathbf{B}_k \cdot \mathbf{\Phi}_z \end{cases} \tag{6.3}$$

where

$$\mathbf{B}_k = [[(B_{i,p}(u_k) \cdot B_{j,q}(v_k))_{j=0}^{n_v-1}]_{i=0}^{n_u-1}] = [\overline{B}_{k,0}, \overline{B}_{k,1}, \ldots, \overline{B}_{k,n-1}]$$

and

$$\mathbf{\Phi}_x = [[(x_{ij})_{j=0}^{n_v-1}]_{i=0}^{n_u-1}]^T,$$

$$\mathbf{\Phi}_y = [[(y_{ij})_{j=0}^{n_v-1}]_{i=0}^{n_u-1}]^T,$$

$$\mathbf{\Phi}_z = [[(z_{ij})_{j=0}^{n_v-1}]_{i=0}^{n_u-1}]^T.$$

If a total of m points on the surface are considered, we have

$$\begin{cases} \mathbf{x} = \mathbf{B} \cdot \mathbf{\Phi}_x \\ \mathbf{y} = \mathbf{B} \cdot \mathbf{\Phi}_y, \\ \mathbf{z} = \mathbf{B} \cdot \mathbf{\Phi}_z \end{cases} \tag{6.4}$$

where $\mathbf{x} = [x_1, x_2, \ldots, x_m]^T$, $\mathbf{y} = [y_1, y_2, \ldots, y_m]^T$ and $\mathbf{z} = [z_1, z_2, \ldots, z_m]^T$. $\mathbf{B}$ is a matrix consisting of the tensor products of the B-spline basis functions corresponding to each of the m measurement points on the surface:

$$\mathbf{B} = \begin{bmatrix} \overline{B}_{0,0} & \overline{B}_{0,1} & \cdots & \overline{B}_{0,n-1} \\ \overline{B}_{1,0} & \overline{B}_{1,1} & \cdots & \overline{B}_{1,n-1} \\ \vdots & \vdots & \cdots & \vdots \\ \overline{B}_{m-1,0} & \overline{B}_{m-1,1} & \cdots & \overline{B}_{m-1,n-1} \end{bmatrix}.$$

If $\mathbf{B}$ in (6.4) is of full rank, then $\mathbf{B}^T\mathbf{B}$ is nonsingular. The least square estimation of $\mathbf{\Phi} = [\mathbf{\Phi}_x^T, \mathbf{\Phi}_y^T, \mathbf{\Phi}_z^T]^T$ can be given as:

$$\begin{cases} \mathbf{\Phi}_x = [\mathbf{B}^T\mathbf{B}]^{-1}\mathbf{B}^T \cdot \mathbf{x} \\ \mathbf{\Phi}_y = [\mathbf{B}^T\mathbf{B}]^{-1}\mathbf{B}^T \cdot \mathbf{y} \\ \mathbf{\Phi}_z = [\mathbf{B}^T\mathbf{B}]^{-1}\mathbf{B}^T \cdot \mathbf{z} \end{cases} \tag{6.5}$$

where $[\mathbf{B}^T\mathbf{B}]^{-1}\mathbf{B}^T$ is the pseudo-inverse matrix of $\mathbf{B}$.

6.2.2 Model Selection

It is known that for a given set of measurement data, there exists a model of optimal complexity that has the smallest prediction/generalization errors for further data. For a B-spline surface, the model complexity is related to the number n ($n = n_u \times n_v$) of control points (parameters) in the u and v directions in the parameter field (Yan et al. 1999). If the B-spline model contains too many control points, the approximated B-spline surface will tend to over-fit noisy measurement data. If the model does not have enough control points, then it will not be able to fit the measurement data, causing the approximation to be under-fitted. In general, both over- and under-fitted approximation will have a poor generalization capability. Therefore, the problem of finding an appropriate model, referred to as model selection, is important for achieving a high level of generalization capability. The problem of model selection has been studied from various standpoints. Examples include information statistics (Sugiyama and Ogawa 2001), Bayesian statistics (Shwartz 1978, Torr 2002) and structural risk minimization (Cherkassky et al. 1999). The Bayesian approach is perhaps the most general and powerful method.

Given a set of models $\{M_k, k = 1,2,\ldots,k_{\max}\}$ and data $\mathbf{r}$, the Bayesian approach selects the model with the largest posterior probability. The posterior probability of model M_k is

$$p(M_k \mid \mathbf{r}) = \frac{p(\mathbf{r} \mid M_k)p(M_k)}{\sum_{L=1}^{k_{\max}} p(\mathbf{r} \mid M_L)p(M_L)} \propto p(M_k \mid \mathbf{r}) \tag{6.6}$$

where $p(\mathbf{r} \mid M_k)$ is the likelihood function of model M_k and $p(M_k)$ is the prior probability of model M_k.

If we assume that the models have the same likelihood a priori, that is $p(M_k) = 1/k_{\max}$, $(k = 1,\ldots,k_{\max})$, the posterior probability $p(M_k \mid \mathbf{r})$ will not be affected by $p(M_k)$. This is also the case with $\sum_{L=1}^{k_{\max}} p(r \mid M_L)p(M_L)$ since it is not a function of M_k. Therefore, the posterior probability $p(M_k \mid \mathbf{r})$ is proportional to $p(\mathbf{r} \mid M_k)$.

To find the model with the largest posterior probability, that is $M = \arg \max_{M_k, k=1,\ldots,k_{\max}} p(M_k \mid \mathbf{r})$, we can evaluate the likelihood function $p(\mathbf{r} \mid M_k)$ of model M_k,

$$M = \arg \max_{M_k, k=1,\ldots k_{\max}} \{p(\mathbf{r} \mid M_k)\} \tag{6.7}$$

To calculate $p(\mathbf{r} \mid M_k)$, we need to calculate multidimensional integration (Torr 2002)

$$p(\mathbf{r} \mid M_k) = \int_{\Phi_k} p(\mathbf{r} \mid \Phi_k, M_k)p(\Phi_k \mid M_k)d\Phi_k$$

In most practical cases, calculating the multidimensional integration is hard, especially to obtain a closed form analytical solution. The research in this area has resulted in many approximation methods for achieving this. Schwarz (1978) and Torr (2002) used Laplace's approximation method for the integration, and simplified $p(\mathbf{r} \mid M_k)$ to

$$\log p(\mathbf{r} \mid M_k) = \log p(\mathbf{r} \mid \hat{\Phi}_k, M_k) - \frac{1}{2}\log | H(\hat{\Phi}_k) |$$

where $\hat{\boldsymbol{\Phi}}_k$ is the maximum likelihood estimate of $\boldsymbol{\Phi}_k$, and $H(\hat{\boldsymbol{\Phi}}_k)$ is the Hessian matrix of $-\log p(\mathbf{r}|\boldsymbol{\Phi}_k, M_k)$ evaluated at $\hat{\boldsymbol{\Phi}}_k$,

$$H(\hat{\boldsymbol{\Phi}}_k) = -\frac{\partial^2 \log p(\mathbf{r}|\boldsymbol{\Phi}_k, M_k)}{\partial \boldsymbol{\Phi}_k \partial \boldsymbol{\Phi}_k^T}\Bigg|_{\boldsymbol{\Phi}_k = \hat{\boldsymbol{\Phi}}_k}.$$

By approximating $\left(\frac{1}{2}\right)\log|H(\hat{\boldsymbol{\Phi}}_k)|$ by the asymptotic expected value $\frac{3\log(m)}{2}d_k$ of Hessian, we can obtain the Bayesian Information Criterion (BIC) (Torr 2002) for selecting the structure of the B-spline surface

$$M = \arg\max_{M_k, k=1,..k_{\max}}\left\{\log p(\mathbf{r}|\hat{\boldsymbol{\Phi}}_k, M_k) - \frac{3\log(m)}{2}\cdot d_k\right\} \tag{6.8}$$

where d_k is the number of control points for B-spline model M_k.

Consider the likelihood function of the parameter of the B-spline model. The probability distribution function $p(\mathbf{r}|\hat{\boldsymbol{\Phi}}_k, M_k)$ of the surface can be factorized into x, y, and z components as

$$p(\mathbf{r}|\hat{\boldsymbol{\Phi}}_k, M_k) = p(x|\hat{\boldsymbol{\Phi}}_{kx}, M_k)\cdot p(y|\hat{\boldsymbol{\Phi}}_{ky}, M_k)\cdot p(z|\hat{\boldsymbol{\Phi}}_{kz}, M_k)$$

Consider the x component. Assuming that the residual error sequence e_{ix} ($e_{ix} = x_i - \mathbf{B}_i\boldsymbol{\Phi}_{kx}$, $i=0,1,\ldots, m-1$) obeys Gaussion distribution with zero mean and variance σ_{kx}^2, the $p(x|\hat{\boldsymbol{\Phi}}_{kx}, M_k)$ can be calculated by

$$p(\mathbf{x}|\hat{\boldsymbol{\Phi}}_{kx}, M_k) = \left(\frac{1}{2\pi\sigma_{kx}^2(\hat{\boldsymbol{\Phi}}_{kx})}\right)^{m/2}\exp\left\{-\frac{1}{2\sigma_{kx}^2(\hat{\boldsymbol{\Phi}}_{kx})}\sum_{i=0}^{m-1}[x_i - \mathbf{B}_i\hat{\boldsymbol{\Phi}}_{kx}]^2\right\} \tag{6.9}$$

with $\sigma_{kx}^2(\hat{\boldsymbol{\Phi}}_{kx}, M_k)$ estimated by

$$\hat{\sigma}_{kx}^2(\hat{\boldsymbol{\Phi}}_{kx}) = \frac{1}{m}\sum_{i=0}^{m-1}[x_i - \mathbf{B}_i\hat{\boldsymbol{\Phi}}_{kx}]^2 \tag{6.10}$$

The $p(y|\hat{\boldsymbol{\Phi}}_{kx}, M_k)$ and $p(z|\hat{\boldsymbol{\Phi}}_{kx}, M_k)$ for y and z components can also be obtained in the similar way. Therefore, we can obtain the following BIC criterion for selecting a B-splines model

$$M = \arg \max_{M_k, k=1,\ldots k_{\max}} \left\{ -\frac{m}{2} \sum_{f=x,y,z} \log \hat{\sigma}_{kf}^2 (\hat{\mathbf{\Phi}}_{kf}) - \frac{3 \cdot \log(m)}{2} \cdot d_k \right\} \qquad (6.11)$$

where m is the number of data points. As the first two terms in (6.11) measure the prediction accuracy of the B-spline model, the BIC criterion will increase as the complexity of the model increases. In contrast, the second term will decrease and act as a penalty for using additional parameters to model the data. However, since the predicted $\hat{\sigma}_{kf}^2$ ($f=x, y, z$) depends only on the training data sampled for model estimation, they are insensitive when under-fitting or over-fitting occurs. In (6.11), only the second term prevents the occurrence of over-fitting. In fact, an honest estimate of σ_{kf}^2 ($f=x, y, z$) should be based on a re-sampling procedure. Here, we can divide the available data into a training sample and a prediction sample. The training sample is used only for model estimation, whereas the prediction sample is used only for estimating the prediction data noise σ_{kf}^2 ($f=x, y, z$). That is, the training sample is used to estimate the model parameter $\hat{\mathbf{\Phi}}_k$ by (6.5), while the prediction sample is used to predict data noise σ_{kf}^2 ($f=x, y, z$) by (6.10). In fact, if the model $\hat{\mathbf{\Phi}}_k$ fitted to the training data is valid, then the estimated variance $\hat{\sigma}_{kf}^2$ ($f=x, y, z$) from the prediction sample should also be a valid estimate of the data noise. If the variance $\hat{\sigma}_{kf}^2$ ($f=x, y, z$) found from the prediction sample becomes unexpectedly large, we have grounds for believing that the candidate model fits the data badly. It is seen that the data noise $\hat{\sigma}_{kf}^2$ ($f=x, y, z$) estimated from the prediction sample is more sensitive to the quality of the model than the one directly estimated from the training sample, as the $\hat{\sigma}_{kf}^2$ ($f=x, y, z$) estimated from the prediction sample also has the capability of detecting the occurrence of under-fitting or over-fitting.

6.3 Uncertainty Analysis

Equation (6.5) produces the parameter estimation of a B-spline model. It should be noted that measurement data are normally contaminated by noise, and it is impossible to find an exact solution for the B-spline model. From here on in this section, we will ignore the k in $\hat{\mathbf{\Phi}}_{kf}$ ($f=x, y, z$) and other symbols related to the selected model M_k for simplification. Since the residual sequence e_f ($f=x, y, z$) obeys Gaussion distribution with zero mean and variance σ_f^2, and the B-spline model in (6.4) is linear, the parameter errors $\mathbf{\Phi}_f - \hat{\mathbf{\Phi}}_f$ are also a Gaussian distribution with zero mean and covariance

$$\mathbf{C} = \sigma^2 \begin{bmatrix} (\mathbf{B}^T\mathbf{B}) & 0 & 0 \\ 0 & (\mathbf{B}^T\mathbf{B}) & 0 \\ 0 & 0 & (\mathbf{B}^T\mathbf{B}) \end{bmatrix}^{-1} = \sigma^2\mathbf{M}^{-1}$$

where $\sigma^2 = \sigma_f^2$ ($f=x$, y, z), which is based on the assumption that the residual sequence e_f ($f=x$, y, z) has the same covariance. Denoting $\mathbf{\Phi} = [\mathbf{\Phi}_x^T, \mathbf{\Phi}_y^T, \mathbf{\Phi}_z^T]^T$, we consider the following quadratic form

$$(\mathbf{\Phi} - \hat{\mathbf{\Phi}})^T \mathbf{C}^{-1}(\mathbf{\Phi} - \hat{\mathbf{\Phi}}) = \frac{1}{\sigma^2}(\mathbf{\Phi} - \hat{\mathbf{\Phi}})^T \mathbf{M}(\mathbf{\Phi} - \hat{\mathbf{\Phi}})$$

that defines the shape of the Gaussian distribution of the parameter error. In fact, the quadratic form defines a hyper-ellipsoid on which the true model parameters must lie. We do not know the position of the shape as we do not know the true value of $\mathbf{\Phi}$. However, we know the range within which the unknown true $\mathbf{\Phi}$ value lies with a confidence interval. For a confidence level γ, we can find from the distribution a number χ_γ^2 for which there is a probability for γ so that $\frac{1}{\sigma^2}(\mathbf{\Phi} - \hat{\mathbf{\Phi}})^T \mathbf{M}(\mathbf{\Phi} - \hat{\mathbf{\Phi}}) < \chi_\gamma^2$. It follows that there is also a probability for γ that yields the hyper-ellipsoid

$$(\mathbf{\Phi} - \hat{\mathbf{\Phi}})^T \mathbf{M}(\mathbf{\Phi} - \hat{\mathbf{\Phi}}) = \sigma^2 \chi_\gamma^2 \qquad (6.12)$$

The true model will be contained in the above ellipsoid which is referred to as the ellipsoid of confidence. The ellipsoid of confidence gives us a useful visual image of the uncertainty of parameter $\mathbf{\Phi}$ of the B-spline surface. In (6.12), $\mathbf{M}$ is also known as the Fisher information matrix (Wang 1999) which characterizes the uncertainty in the estimated parameters. Therefore, the problem of selecting an optimal set of measurement data for CMM's high-precision measurement is to find the locations of the measurement data points for which the estimation uncertainty is minimized in some sense. Various criteria exist for optimizing the Fisher information matrix to achieve minimum estimation errors. The major criterions include $Cond(\mathbf{M})$, $Trace(\mathbf{M})$ (A-optimality), the maximum eigenvalue of $\mathbf{M}^{-1}$ (E-optimality), and $|\mathbf{M}|$ (D-optimality) (Wang 1999, Chio and Kurfess 1995). These criteria measure the amount of information contained in the probability distribution representing the parameter errors. Thus, ensuring that the important information and necessary information in the B-spline model is embodied in the measurement data set is the primary concern in selecting an optimal set of measurement data for CMM's high-precision measurement. Here the optimal criterion adopted is the D-optimality, or the determinant criterion, for which the determinant of the Fisher information matrix $|\mathbf{M}|$ is to be maximized. Geometrically, the volume of the ellipsoid is inversely proportional to the square root of the determinant $|\mathbf{M}|$. A large $|\mathbf{M}|$ corresponds to a small volume of the model parameter space, indicating that the true parameters are well localized and that the knowledge or information we have about them is highly reliable (Wang 1999, Chio 1995, Whaite 1997). Here, we define $|\mathbf{M}|$ as the uncertainty measurement for the estimated parameter vector $\mathbf{\Phi}$.

6.4 Sensing Strategy for Optimizing Measurement

As the uncertainty of a B-spline model is dependent on the number and locations as well as the variance of the measurement data, the sensing strategy plays a critical role in the measurement and reconstruction results. A sensing strategy should be able to determine the number of measurement data to sample and the locations to take the measurements, while keeping the uncertainty of the reconstructed B-spline model sufficiently low.

6.4.1 Determining the Number of Measurement Data

Since the reconstruction of a freeform surface is based on the measurements at discrete points to be sensed by a CMM's touch probe, these discrete points must contain sufficient information that allows the freeform surface to be reconstructed. However, the number of measurement data has to be limited to achieve a reasonable speed in the measurement process. From the statistical point of view, the number of measurement data should be at least ten times the number of the parameters in the B-spline model to make the B-spline regression analysis statistically meaningful (Yang and Menq 1993). For example, for a bi-cubic B-spline surface with d ($d = n_u \times n_v$) control points, at least $10 \times d$ measurement data are required.

6.4.2 Optimizing the Locations of Measurement Data

Since $|\mathbf{M}|$ is dependent not only on the number of measurement data, but also on the locations of the measurement data, we should also optimize the locations of the measurement data to maximize $|\mathbf{M}|$. The parameter variables u and v of the measurement data in the parameter field of the freeform surface constitute the design variables. Each candidate measurement data point can vary its location (u, v) within a specified range. The coordinate (x, y, z) of the measurement data can be obtained from the parameter variables (u, v) by (6.4). Thus, optimizing the locations of the measurement data points for minimizing the uncertainty of a B-spline model can be stated as follows:

$$\max_{u_k, v_k} |\mathbf{M}| \tag{6.13}$$

subject to: $(u_i, v_i) \in [0,1]$, $i = 0,1,...m-1$.

The problem is essentially a combinatory optimization problem. Since the objective function $|\mathbf{M}|$ is non-smooth and nonlinear, the existence of the derivations at all points is not guaranteed. This makes the optimization difficult if a standard optimization method is used. To simplify the problem, $|\mathbf{M}|$ can be evaluated with an existing discrete D-optimal design method called Fedorov exchange algorithm (Miller and Nguyen 1994). This algorithm implements an efficient neighborhood search for the maximum determinant of the Fisher information matrix $\mathbf{M}$.

Consider the incremental form of $|\mathbf{M}|$. Each additional measurement data incrementally updates $\mathbf{M}$, so that after $i+1$ measurements, its value becomes $\mathbf{M}(i+1) = \mathbf{M}(i) + L_{i+1}^T L_{i+1}$. The corresponding determinant of $\mathbf{M}$ then is

$$|\mathbf{M}(k+1)| = (1 + L_{i+1} \cdot \mathbf{M}^{-1}(k) \cdot L_{i+1}^T) \cdot |\mathbf{M}(k)| \qquad (6.14)$$

where $L_{i+1} = [\mathbf{B}_{i+1}, \mathbf{B}_{i+1}, \mathbf{B}_{i+1}]$, $\mathbf{B}_{i+1}$ is the basis function vector evaluated at location (u_{i+1}, v_{i+1}).

If a point is to be removed from the set of sample points, all the plus and minus signs in (6.14) are reversed. To evaluate $|\mathbf{M}|$ by Fedorov exchange algorithm, each point in the set of measurement data is considered for exchange with each of the available candidate points. The pair of points chosen for exchange is the pair that maximizes the increase in the determinant of $\mathbf{M}$. This process is repeated until no further increase in the determinant can be obtained by the exchange.

If we denote the point to be added by L_+, and the point to be replaced by L_-, then by exchanging the pair of L_+ and L_-, the new determinant is

$$|\mathbf{M} + L_+^T L_+ - L_-^T L_-| = |\mathbf{M}| \cdot [1 + \Delta(L_+, L_-)] \qquad (6.15)$$

where

$$\Delta(L_+, L_-) = L_+ \mathbf{M}^{-1} L_+^T - L_- \mathbf{M}^{-1} L_-^T (1 + L_+ \mathbf{M}^{-1} L_+^T) + (L_+ \mathbf{M}^{-1} L_-^T)^2. \qquad (6.16)$$

It is obvious from (6.16) and (6.16) that it is critical for the Fedorov exchange algorithm to find a candidate point to replace a point in the current measurement data set in turn, which maximizes $\Delta(L_+, L_-)$. In this work, we used a simulated annealing algorithm to search the candidate point. Simulated annealing (SA) is a random search algorithm that is popular for solving both the continuous and the discrete global optimization problem. The optimal procedure using the discrete SA algorithm for optimization of the locations of the measurement data points can be stated briefly as follows:

- **Step 1** Select a measurement point $r(u_i, v_i) \in S$, $i = 0, 1, \ldots, m-1$ from the set of sample points.
- **Step 2** Generate a candidate point $r_c(u_c, v_c) \in S$ according to a specified generator.

- **Step 3** Set

$$r_i(u_i,v_i) = \begin{cases} r_c(u_c,v_c) & if\ \Delta(L_+,L_-) > 0 \\ r_c(u_c,v_c) & \text{with probability } p \text{ if } \Delta(L_+,L_-) < 0 \\ r(u_i,v_i) & \text{otherwise} \end{cases}$$

where p is the probability of accepting p when $\Delta(L_+, L_-) < 0$. For simplicity, the probability p is set as constant.
- **Step 4** Repeat Step 2 and 3 until the stopping criterion is satisfied.
- **Step 5** Select another measurement data point from the sample set, and repeat step 1–4 until all measurement data in the selected measurement are exchanged.

6.5 Experiments

To demonstrate the effectiveness of the proposed sensor planning strategy for reconstructing freeform surfaces, experiments are conducted. In the current implementation, a uniform cubic B-spline model is used to represent these surfaces.

One example is an object manufactured in our own laboratory. This object has a freeform surface contained in an area of 40×40 mm^2 and a depth of 10 mm, as shown in Fig. 6.2a.

To reconstruct the freeform surface, the first thing is to determine the control point number n_u and n_v of the B-spline model in the u and v parameter directions. A 3D vision system was used to acquire a cloud of data points on an object surface. This vision system (Kreon/KLS51 by Kreon Technologies) consisting of a laser stripe projector and CCD camera measures 3D coordinates based triangulation. The measurement for the above example object is shown in Fig. 6.2b. We used our Modified BIC criterion to select the B-spline model structure (n_u and n_v) for

 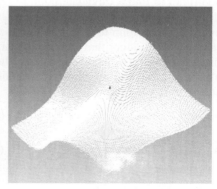

(**a**) The object with a freeform surface (**b**) The point cloud acquired by a 3D vision sensor

Fig. 6.2. The experimental object

representing the freeform surface. To demonstrate the effectiveness of the modified BIC criterion, we compared it with the BIC and cross validation (CV) methods (Cherkassky et al. 1999, Mcquarrie and Tasi 1998) respectively. The two following performance indexes were used:

1. Estimation accuracy, which is defined as the MSE (Mean Square Error) between the actual data points and the regression estimate chosen by a given model selection method;
2. Model complexity, which refers to the number d ($d = n_u \times n_v$) of control points of a B-spline model determined by a given model selection criterion.

In this section, we use box plots of the MSE and model complexity of each method to test the performance of different model selection methods. The experiments with different sample sizes were designed to observe the differences between the different model selection methods. For each sample size, the sample points were selected randomly from the "data cloud" acquired by the 3D vision system, and then used to determine the model structured out of the B-spline model with a different model selection criterion. The above selection process was repeated 100 times. The comparison results are presented in Box plots which give the empirical distribution of the comparison based on 100 iterations in the model selection. An evaluation result with a set of 300 sample points is shown as a box plot in Fig. 6.3. In this figure, the box represents the range of distribution of the quantity under study. The box stretches from the lower hinge (defined as the 25th percentile) to the upper hinge (the 75th percentile) and therefore contains the middle half of the scores in the distribution. The dark line (shown as across a box) is the median of the quantity. Therefore ¼ of the distribution of a box lies between this dark line and the top of the box, and ¼ of the distribution lies between this dark line and the bottom of the box.

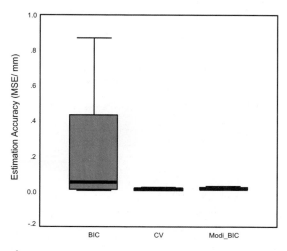

Fig. 6.3. Estimation Accuracy

The MSE box plot, in combination with the box plot of model complexity (i.e. the number of control points), provides visual judgment on the relative performance of various model selection methods. A lower value of MSE in the plot corresponds to a better model selection approach. The model complexity plot, together with the estimation accuracy plot, provides information on the over-fitting or under-fitting for a given method relative to the optimally chosen model complexity. The height of the bar in the plots of the estimation accuracy reflects the method's sensitivity to random sample variations, which can be used as a measure of the variability in the error estimation. A short bar in the plot indicates that the method is insensitive (robust) to random variations in the data. In general, low model complexity is desired. As the number of parameters in a B-spline model is related to its uncertainty, the more the parameters of a B-spline model, the higher the uncertainty tends to be. For a model with high complexity, more measurement data would be needed to increase the reliability in the parameter estimation. In such a case, the time cost in the measurement and reconstruction would be high.

From Figs. 6.3 and 6.4, the model selected by BIC provides a consistent model structure, which is insensitive to random variances in the data. However, the estimation accuracy is rather poor, compared with the CV and our modified BIC method (denoted as Modi_BIC) as can be seen in Fig. 6.3. In fact, in the BIC criterion, only the second term can prevent over-fitting. As a result, BIC is insensitive to over-fitting and a model with high complexity is selected. On the other hand, the re-sampling procedure in CV and our method has the capability of detecting the occurrence of over-fitting and under-fitting in time. Compared with CV, our criterion results in a similar level of estimation accuracy and provides a

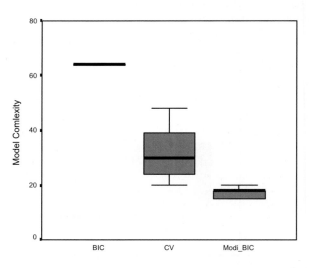

Fig. 6.4. Model complexity (Results with 300 training samples and 200 prediction samples)

lower complexity model with satisfactory consistence. We also compared the performance of the three model selection methods using a larger sample size of 1200. It was observed that the models selected by the three methods had similar levels of estimation accuracy (about 0.012 mm), while the BIC and CV method seem to prefer a model with higher complexity (with the median of 80 for BIC, 69 for CV, and 33 for our modified BIC), although BIC exhibited good consistency and insensitivity to random variances. With our criterion, a much lower model complexity was achieved, while keeping similar estimation accuracy. On the whole, our method achieved a good compromise between the selected model complexity and estimation accuracy.

Then we further tested our method (modified BIC criterion) with different sized samples, where the number of prediction samples used was about 40% of that of the training samples. The results are given in Table 6.1. It can be seen that with the increase in the sample size, the estimation accuracy tends to improve while the model complexity tends to increase. Such an effect becomes less obvious when the sample size is bigger than 1200, where the model complexity and accuracy tend to be stabilized. In such a case, the corresponding model structure can be considered as converged to the true model of the freeform surface. In our system, since we can get a sample set with a sufficiently large size from a cloud of data obtained by the vision system, we can assume that the true model structure to describe the unknown freeform surface can be obtained.

Here, we used bi-cubic B-splines to model the freeform surface. Different B-spline models with different control points in u and v directions were evaluated by our modified BIC criterion. The result was a B-spline model with 6 control points in both u and v directions respectively (totally 36 parameters to be estimated) which had the highest scores of our modified BIC. This model is a result yielded by our method to represent the freeform surface to be reconstructed.

Based on the selected B-spline model, the minimal set of 360 measurement data was used to estimate these parameters. As discussed in Sect. 3, high uncertainty in the estimated parameters indicates that the estimated values of $\hat{\Phi}$ can deviate significantly from the true values of Φ. In other words, the lower the uncertainty in the estimated parameters, the more reliable the estimation $\hat{\Phi}$ is. Here, we use the $\log(|M|)$ as the indicator of the uncertainty in a B-spline model. The larger the $\log(|M|)$, the lower the uncertainty.

Table 6.1. Results of model selection by modified BIC with different sample sizes

Sample size	200	300	500	800	1200	1600	2000	2500
Accuracy (MSE)	0.075	0.019	0.022	0.019	0.015	0.015	0.015	0.015
Model Complexity	2×3	4×4	4×4	4×5	6×6	6×6	6×6	6×6

Next, we employed the Fedorov exchange algorithm to optimize the locations of the measurement data. The locations of the measurement data before and after the optimization are shown in Figs. 6.5 and 6.6. Before optimization, the measurement data were located randomly in the parameter space (u, v) of the B-spline surface, with the uncertainty of the B-spline model $\log(|\mathbf{M}|)$ being -120. Using the Fedorov exchange algorithm, the locations of sample points were adjusted one by one, with the $\log(|\mathbf{M}|)$ value of the B-spline model increased gradually to -94.5, which shows a significant decrease in the corresponding uncertainty of the B-spline compared with using random locations in the measurement data. On the other hand, increasing the sample size can also reduce the uncertainty of a B-spline model. To achieve the same level of uncertainty in the B-spline model with random locations in the

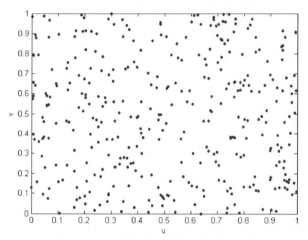

Fig. 6.5. The locations of the measurement data before optimization

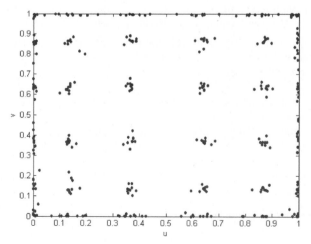

Fig. 6.6. The locations of the measurement data after optimization

measurement data, about 470 more measurement data would be needed in the sample set. This shows that optimizing the locations of the measurement data point to be sensed by CMM's touch probe can yield much more reliable model estimation, without increasing the number of measurements to be taken. We also compared our optimization results with a measurement of equidistant probing points. The uncertainty $\log(|\mathbf{M}|)$ of the B-spline model using equidistant probing was found to be -220.1 which is much worse than our optimization result.

Here an interesting phenomenon to note concerning the optimized locations of the measurement data is that after optimization, the measurement data are located in the neighborhood of each model parameter. These relocations allow for a more reliable model estimation in the parameterization space. The coordinates (x, y, z) of the measurement data can be mapped from the parameter variables (u, v) with appropriate coordinate transformations. Finally, the surface of the object in Fig. 6.2, reconstructed using our method, is shown in Fig. 6.7. Here, the mean deviation of the measured coordinates from the reconstructed surface is 0.012 mm, while the minimum deviation is 0.0011 mm and the maximum deviation is 0.028 mm.

From the experiments, we observed that in the parameter space, the locations of the measurement data points are related to the structure of the B-spline model. For a uniform cubic B-spline model, the control points are distributed uniformly in the u and v directions, giving rise to some clusters in which the measurement data points are located. Therefore, we infer that the structure of a B-spline model determines the locations of the measurements and the model structure represents the geometrical feature of a surface, which can be extracted from the cloud of data acquired by a vision system. In addition, as there is a coordinate transformation

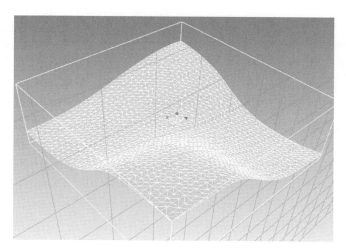

Fig. 6.7. The reconstructed freeform surface

between the location parameters (u, v) and the corresponding spatial coordinates (x, y, z), different parameterization methods can influence the optimization results. We further infer that if we locally modify the distribution of the control points according to the geometrical feature of a surface, the distribution of the measurement data will be changed accordingly.

6.6 Summary

In this chapter, we present a sensing strategy for optimal measurements for the reconstruction of freeform surfaces. We assume the availability of a vision system to quickly obtain the rough data of a surface for guiding the more accurate but much slower touch sensing such as the touch probing in CMM. We investigated the use of B-spline models to represent freeform surfaces and proposed the modified BIC method for selecting the optimal model structure from the cloud of data points acquired by a 3D vision system. Based on the model structure, the number of measurement data needed for the high-precision measurement is then determined. In order to obtain a more accurate model, the uncertainty of the model is analyzed. Then using the statistical analysis of the Fisher information matrix, the locations of the measurement data points are optimized to reduce the uncertainty in the model. Based on the results of the optimized measurements, a more accurate touch sensing, e.g. by CMM's touch probe, can be used to obtain the accurate measurements for the reconstruction of the freeform surface more efficiently. The proposed method will allow the advantages of the high speed in vision sensing and the high accuracy in touch sensing to be utilized for efficient and accurate reconstruction of freeform surfaces. The experimental results show that the proposed method is effective and promises useful applications in integrated multi-sensor measurements such as vision-guided CMM for reverse engineering. When combined with an adaptive modeling scheme based on the features of a freeform surface, adaptive localization of the measurement data points can also be implemented.

Chapter 7
Sensor Planning for Object Modeling

While sensor placement for model-based vision tasks has been discussed in the previous chapters, we are about to present a method of viewpoint planning for incrementally building the model of unknown objects or environments by an active visual system in the following three chapters. We firstly list some typical approaches to sensor planning for model construction, including the multi-view strategy and existing contributions. The standard procedure for modeling of unknown targets is provided. A self-termination judgment method is suggested based on Gauss' Theorem by checking the variations of the surface integrals between two successive viewpoints so that the system knows when the target model is complete and it is necessary to stop the modeling procedure.

7.1 Planning Approaches to Model Construction

7.1.1 Model Construction from Multiple Views

Multiple views are required to reconstruct a 3D model of a complete object or environment. Some single depth images are acquired from several views and merged together with geometric fusion techniques to produce a representation of the underlying 3D target. This is the basic idea in model-building tasks.

Considerable relevant works have been conducted recently for constructing the model of a scene or building. Gimel'farb and Haralick (1997) conducted experiments in voxel-based multiple-view terrain reconstruction. They described the reconstruction of the RADIUS model-board scene from 40 pre-calibrated images collected by different cameras having different positions and orientations, under various illumination, and at different times. Bolter and Leberl (2000) studied the detection and reconstruction of buildings. The exploitation of the building becomes feasible by combining multiple views and multiple data types of the same scene. They used information from the interferometric height and coherence data to separate regions containing buildings from other objects in the scene. Shadow information from magnitude images was then used to delimit the exact boundaries of the buildings further. Liebowitz and Zisserman (1999) also presented an approach to combining scene and auto-calibration constraints. Calibration constraints were provided by imaged scene structure, such as vanishing points of orthogonal directions, or rectified planes.

Before multiple views are integrated to form a complete 3D model, a registration step is necessary to find the transformation matrix between each pair of views. The view registration is to align the point sets of the different views, so that errors in the overlapping regions are minimized. If two sets of corresponding points have been identified, registering two range images can be done using a quaternion-based non-linear optimization method as described in Horn (1987), Besl and Mckay (1992). For integration of both range and intensity information in the registration process, Weik (1997) proposed to use the intensity information for point correspondences. Lucchese et al. (1997) proposed a method based on 3D Fourier transform, in which the intensity information is used only for the disambiguation of the shape-based registration results. Pottmann et al. (2002) presented an iterative algorithm which simultaneously registers all 3D image views. Ho and Chua (1999) presented to register surfaces using Point Signatures. Multiple range images were extracted from various unknown viewpoints and integrated to form a complete 3D representation of the model. A long list of other works on registration can also be found in the literature but will not be given here as it is not the focus of this research.

The problem of 3D model construction was also addressed in Chen et al. (1999) for a free-form object from multiple range images using spherical harmonics. The relative phase of each pixel on the object surface was computed from five phase-stepped interferometry images. Phase unwrapping, which converts relative phase into absolute phase, was then applied to obtain a range view of the sensed 3D surface. Multiple range views were first aligned with one another and then integrated to form a 3D object model. Higuchi (1995) described an approach to building a 3D model from a set of arbitrary range images without initial estimate of the relative viewpoints. The approach was based on building discrete meshes representing the surfaces observed in each of the range images, mapping each of the meshes to a spherical image, and computing the transformations between the views by matching the spherical images.

On direct surface reconstruction, recently Siddiqui and Sclaroff (2001) proposed a method for reconstruction of 3D rational B-spline surfaces from multiple views. Given corresponding features in multiple views, the surface was reconstructed. After 2D B-spline patches were fitted to each view, the 3D B-splines and projection matrices could then be extracted from the 2D B-splines using factorization methods. The surface fit was further refined via an iterative procedure. A hierarchal fitting scheme was proposed to allow modeling of complex surfaces by means of knot insertion. In a more complex method, (Shapiro 1995) presented a domain-model approach to reconstruction of 3D environments for virtual reality. The approach was to use domain knowledge to simplify and improve the model-acquisition process. It was knowledge-driven and attempted to understand the physical structure of the environment and of the individual objects in the environment. They proposed to develop physical domain models that would define the physical constraints of a particular domain. Such models would include such information as possible 3D surface classes, possible materials, surface relationships, common 3D primitive solids, functional relationships among the 3D primitives, primitives, and fixed or constrained lighting and sensor information.

Gray et al. (2001) presented a model creation using multiple range and intensity image pairs of an object. All of the pairs were assumed to have been registered to a global coordinate system. The individual range images were used to create a surface mesh and the associated intensity images were applied to the surface mesh as a texture map. Some general representations of target surfaces have been considered in a bottom-up framework, where many small planar regions were fitted to an object. In particular, planar meshes were explored (Fua and Leclerc 1996). They further modeled the surfaces as oriented particles or tiny planes with associated texture (Fua 1997). While generally in practice, many facets are needed for these methods to successfully approximate a surface. In many cases objects intrinsically do not have very high complexity and more appropriate classes of models are sought. Shashua and Toelg (1997) proposed a method that can be used when objects are well approximated by quadric surfaces by examining the induced flow field. Lin et al. (2002) also proposed an approach to reconstructing an environment model by using a well-calibrated active binocular head. The reconstructed 3D points and their gray level values are stored in the inverse polar octree. An active control scheme has been used to minimize the ambiguity in stereo matching.

To reconstruct a complete model of an unknown object, for which no priori information is available, it is insufficient to simply fuse several views from un-carefully planed positions. It is necessary to locate sensors in the environment strategically, since not all surfaces may be visible from a single point, nor will data be acquired at sufficient resolution. Mobility or sensor placement is therefore paramount to the 3D reconstruction. Sequeira et al. (1996, 1999) proposed the perception planning algorithms recently. The algorithms start by detecting the occlusions on the current 3D reconstructed environment, followed by the evaluation of the set of potential capture points from which all the occlusions can be resolved. This set is fed into an optimization procedure aiming at: (1) minimizing the number of capture points; (2) selecting those points from which the occlusions areas are captured, as much as possible, along the normal to the occluded plane, (3) selecting those points from which the distance to the occlusions leads to smaller errors on range acquisition, i.e., minimize depth error.

An intensive survey of object reconstruction with view planning leads to the problem of *sensor placement for observing unknown objects*, which has been presented in the previous chapter. The author proposed an incremental method for complete object reconstruction. The next view is determined according to the trend surface which is proposed as the cue to predict the unknown portion of an object or environment. This method is target-driven and best suited for modeling large continuous objects or environments. The following sections present a cue to derive real dimensions and provide some experiments of surface reconstruction and complete object reconstruction.

7.1.2 Previous Planning Approaches for Modeling

In some vision tasks of modeling and autonomous exploration, there is often no priori information provided, and even nearly nothing is known about the object's geometry except that it has a certain extent.

Since reconstruction of a model for a complete object surface requires images from multiple views (Banta 1996) and prior to data acquisition the number of views for an unknown object and their optimal positions is not known, techniques are required to select the *next best view* (NBV) based on the measurement of part of an object's surface. The optimal set of views for capturing a complete object surface will depend on both the unknown surface shape and the known sensor geometry and degrees-of-freedom.

In general, the following three steps are necessary to build a complete surface description of an object by using 3D images (Zha et al. 1998): (1) acquiring range images from different viewpoints; (2) registering the images into a common object-centered coordinate system; (3) integrating the range views into a non-redundant model description. In primitive object reconstruction and depth estimation tasks using mobile intensity cameras, the problem is to choose camera motions which minimize error in the parameter estimation algorithms. In surface reconstruction tasks using intensity cameras (Kutulakos et al. 1994), the problem is to control the camera's motion to guarantee local and hence global surface reconstruction. Finally, in scene reconstruction, the task is to build a model of an unknown scene, perhaps for path planning and potentially with unknown extent.

The majority of work carried out in this sensor position planning is mainly concerned with finding the best views to digitize an object without missing zones, and with a minimum number of views. Varying the view parameters causes the observed features to undergo measurable local transformations which can be used to simplify and constrain the computation of unknown scene parameters. Kutulakos and Dyer (1994) exploited the differential properties of smooth surfaces to model local changes in the appearance of an occluding contour due to camera movement. This knowledge shows them how to position the camera, first to extract occluding contours from an edge map, and then to use the extracted contours to sweep out the complete 3D shape.

With the visibility constraints, this sensor placement problem is addressed of deciding which areas of the viewing volume need to be scanned by identifying discontinuities either in each 3D image or in the model under construction (Maver et al. 1993, Zha 1997, Pito 1999). Compared with Ahuja and Veenstra (1989), who have considered the problem of the views needed to build an octree representation of a 3D scene, Banta et al. used uniformly sized voxels also tagged as either empty or not, to represent the viewing volume. The next best view was identified as the one that would sample the most nonempty voxels. Papadopoulos-Orfanos and Schmitt (1997) also utilized a volumetric representation, but concentrated on a solution to the next best view problem for a short field-of-view range scanner. Their work focused on collision avoidance because the reason of their small field of view causes the sensors to navigate closely to the object. The system digitizes n views, separated by an angle of $2\pi/n$, where n is a parameter chosen by the operator. Tarabanis and Tsai (1991) have worked on a theoretical analysis of the best camera viewpoint for detecting a generic feature.

Occlusion has been strongly associated with viewpoint planning in the modeling research literature for some time. Kutulakos et al. (1994) utilized the changes in the boundary between sensed surface and occlusion with respect to sensor position to recover shape. A similar histogram-based technique was used by Maver and Bajcsy (1993) to find the viewing vector that would illuminate the most edge features derived from occluded regions. Whaite and Ferrie (1997) used a sensor model to evaluate the efficacy of the imaging process over a set of discrete orientations by ray-casting: the sensor orientation that would hypothetically best improve the model is selected for the next view. The work by Pito (1999) removed the need to ray-cast from every possible sensor location by determining a subset of positions that would improve the current model. Pito (1999) presented a solution for the next best view problem (NBV) of a depth camera in the process of digitizing unknown parts. The system builds a surface model by incrementally adding range data to a partial model until the entire object has been scanned. No assumptions are made about the geometry or topology of the object.

Thus NBV planning algorithm is an incremental model construction method composed of a number of observing-and-planning loops. Based on a partial model created thus far, this algorithm provides quantitative evaluations on the suitability of remaining viewpoints as the NBV. Zha et al. (1998) addressed two main issues to determine the next-best-viewpoint: (1) a uniform tessellation of the spherical space and its mapping onto the 2D array; (2) incremental updating computations from evaluating viewpoints as the NBV. They represented the un-scanned areas of the viewing volume with vectors "attached" to the boundaries of surface meshes, which is for creating a complete model of a curved object from multiple range images. Yu et al. (1996) proposed to determine the next pose of the range sensor by analyzing the intersections of planar surfaces obtained from the previous images, so that the unseen parts of the scene can be observed most. This does not depend on a priori geometrical information about the scene. They set up a sphere around the scene and the next view is selected for getting the largest unseen area. Arbel et al. (1999) showed how entropy maps can be used to guide an active observer along an optimal trajectory and how a gaze-planning strategy can be formulated by using entropy minimization as a basis for choosing a next best view.

In 1996, Banta and Abidi described a system to automatically determine an optimized next range sensor position and orientation during the reconstruction of a three-dimensional model. The developed system reconstructs a model consisting of surfaces which have been viewed and volumes occluded from the camera's view. Ideally, a sensor pose determined by a "best-next-view" system will reveal the greatest quantity of previously unknown scene information. The algorithm attempts to intelligently cluster the occluded data and orient the sensor on the centroid of the largest cluster.

Maver and Bajcsy (1993) proposed to solve the next best view problem for a specific scanner and scanning setup consisting of an active optical range scanner and a turntable which rotated between scans. The unseen portions of the viewing volume were modeled as 2.5D polygons where each edge of a polygon was given a height corresponding to the median of the heights of its pixels. The idea is to scan into these polygons from directions which are not occluded. The solid angles in [0, 2pi] which have an unobstructed line of sight into some un-scanned portion of the viewing volume were accumulated in a histogram and the next best view was chosen as the angle with the largest value. In a similar but more general solution, Reed and Allen (1999) determined the visibility volume, which is the volume of space within which a sensor has an unobstructed view of a particular target.

For scene exploration, Birnbaum et al. (1993) presented a system which uses a generative causal semantics to control the visual exploration of arbitrarily stacked block structures. The system works by encoding the knowledge that the scene is stable under the force of gravity in a simple set of rules. The rules are used to direct visual attention to search for blocks which make an unstable stable. Marchand et al. (1999) also dealt with the 3D structure estimation and exploration of a scene using active vision (Bajcsy 1988), whose purpose is handled at two levels: a local aspect where active vision is used to constrain the camera motion in order to improve the quality of the reconstruction results, and a global aspect which is used to explore the unknown areas. The scene is assumed to be only composed of polyhedral objects and cylinders. The technique proposed to solve the "next best view" problem is a depth-first search algorithm and the strategy ensures the completeness of the scene reconstruction.

As presented in Chap. 1, the fundamental objective of sensor placement in such nonmodel-based vision tasks is to increase knowledge about the unseen portions of the viewing volume while satisfying all the placement constraints such as in-focus, field-of-view, occlusion, collision, etc. The previous research efforts were often concentrated on finding the best next views by volumetric analysis or occlusion as a guide. However, since there does not exist any information about the unknown target, it is actually impossible to give the true best next view. In this work, the method involves the decision of the exploration direction and the determination of the best next view and the corresponding sensor settings. The trend surface is proposed as the cue to predict the unknown portion of an object or environment and the next best viewpoint is determined by the expected surface. The viewpoint determined in such a way is predictably best. These works will be further discussed in detail in Chaps. 8 and 9.

7.2 The Procedure for Model Construction

The modeling process is to explore and obtain the unknown portion of the object/environment, in which an incremental method is very common. It results in a sequence of viewing operations for local model acquisition. Each successive sensing result with new information then is merged with the global model being built by proper registration.

During such a process, the first view is obtained at an arbitrary pose, a 3D depth map M_0 is created, and then it is registered as the initial object model O_0. At the same time, the sensor pose is recorded as P_0. Then a next view pose P_{i+1} ($i=0, 1, ...,$ n) must be determined so that the modeling process can continue until the whole object is reconstructed.

Figure 7.1 illustrates five viewpoints needed for acquisition of an object geometry, from P_1 to P_5. Since there is no prior information about the object, each viewpoint (from step 2 to 5) must be decided in run-time. The key problem in this task is to plan a most feasible viewpoint, called the Next-Best-Pose (Zha 1997) or Next-Best-View (Wong et al. 1999), for performing the successive vision perception, according to the partial model that has already been acquired. A next chapter is going to propose a strategy based on the trend surface for a viewpoint decision. Algorithms are developed to dynamically determine the sensor's position, orientation, and optical settings.

Generally, an entire object/environment model is constructed in three stages. First, the 3D sensing technique is applied to measure the shapes of the objects in the scene. Then these local shape models are integrated into a single global model to obtain the complete shape. Finally the shape model can be rendered.

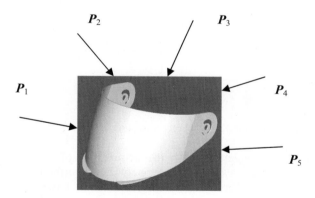

Fig. 7.1. Multiple viewpoints for incremental modeling

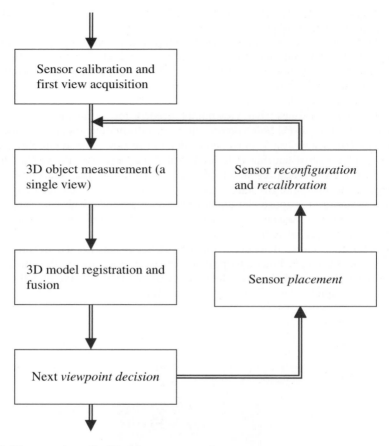

Fig. 7.2. The procedures for 3D object reconstruction

If we are only concerned with the acquisition of the object model, the reconstruction process can more precisely be described with six steps, as illustrated in Fig. 7.2. First, the vision sensor must have been carefully calibrated in advance. A first view is captured for scene analysis. Then, a partial model of the target is acquired by a view setting. A 3D reconstruction method is applied in this step for converting the 2D images into 3D information. Third, the 3D local model is registered and fused with the global model. Here the model is checked to test if it is already finished. If not, a new viewpoint need be decided. The new viewpoint should be "best" so that the whole modeling task can be finished with the highest possible efficiency and accuracy. If a new viewpoint is decided (with methods to be proposed in a later chapter), the sensor will be moved to the corresponding pose (position and orientation) with proper settings. Since the sensor's setting (configuration) may be changed, it has to be calibrated by a self-recalibration method. Then the modeling cycle continues in the next view acquisition and model construction.

7.3 Self-Termination Criteria

To automatically construct a 3D model, a primary consideration is that of determining when planning is no longer needed and the modeling process is complete. The entire process consists of 5 repeated steps: determine the next best view and move the object/sensor system, acquire the object again, register the new range data and integrate the new range data with the partial surface, self-terminate the procedure of acquiring surface data.

On self-termination criteria, Banta and Abida (1996) used a surface area at each new viewpoint. The system would terminate the reconstruction if the ratio of surface faces to occluded faces is large and the change in the surface face count is small or if the change in the occluded face count is small. Arbel and Ferrie (1999) used the entropy value in the termination condition. The system iteratively measures the entropy until it reaches a small enough convergence value.

None of the above criteria are related to explicit requirements to be met by the reconstructed object model, and the threshold value used for the termination judgment can badly affect the measurements. Furthermore, the termination criteria only use measurement data from one viewpoint for the termination condition, whereas the previous measurement data are not taken into account. As a result, these termination criteria are not robust, especially when dealing with complex objects.

In this section, we present a method that can automatically and efficiently acquire the 3D model of an object with self-termination, by calculating the volume encompassed by data from the available views and analyzing the variation in the volume in two successive viewpoints.

7.3.1 The Principle

Suppose we have a surface normal integral over a closed surface S,

$$I = \oint_S f(r) \bullet \mathrm{d}S,$$

and this surface is the surface of a volume V. The surface normal points outwards from the enclosed volume. From Gauss theorem, the integral then becomes the flux of the volume V (see the figure below).

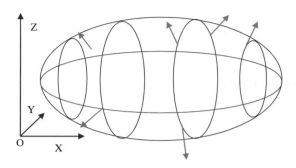

Fig. 7.3. The principle of Gauss' theorem

Here, if

$$f(r) = x \cdot \mathbf{i} + y \cdot \mathbf{j} + z \cdot \mathbf{k}, \tag{7.1}$$

then

$$I = \oint_V dV = \frac{1}{3} \oint_S f(r) dS = \frac{1}{3} \oint_S x dx dy + y dz dx + z dx dy \tag{7.2}$$

Following the measurements from the planned views, the volume encompassed by data points of the object surface will be convergent to a steady value. If this condition is satisfied, then the measurement can be terminated. In this way, the self-termination condition is converted to calculating the volume of the object. As the shape of the object surface can be arbitrary, it is difficult to calculate the object volume directly. Here, we will use a surface integral to replace the volume integral based on Gauss' theorem (Fig. 7.3).

7.3.2 Termination Judgment

As mentioned above, the volume encompassed by data points can be obtained by computing the surface integral. Here we present a method for computing surface integrals using triangular meshes.

For a triangular mesh Δ_i $i = (1, 2, \cdots, n-1, n)$, we can obtain its plane equation equation,

$$a_i x + b_i y + c_i z + d_i = 0 \tag{7.3}$$

The normal vector of this triangular mesh plane is

$$\vec{n} = \frac{1}{\sqrt{a_i^2 + b_i^2 + c_i^2}} (a_i, b_i, c_i) = (\alpha_i, \beta_i, \gamma_i) \tag{7.4}$$

Then, the volume integral can be defined as

$$V = \iiint_V dv = \sum_i^n \iint_{\Delta_i} z dx dy + y dx dz + x dy dz \tag{7.5}$$

For such a triangular mesh, its surface is

$$\begin{aligned} V_i &= \iint_{\Delta_i} z dx dy + y dx dz + x dy dz \\ &= \iint_{\Delta_i} \frac{a_i x}{\sqrt{a_i^2 + b_i^2 + c_i^2}} + \frac{b_i x}{\sqrt{a_i^2 + b_i^2 + c_i^2}} + \frac{c_i x}{\sqrt{a_i^2 + b_i^2 + c_i^2}} ds \\ &= \iint_{\Delta_i} (\alpha_i x + \beta_i y + \gamma_i z) ds \end{aligned} \tag{7.6}$$

Equation (7.6) is a vector function. To simplify its computation, we change the vector function surface integral to a scalar function via

$$\begin{aligned} V_i &= \iint_{\Delta_i} (\alpha_i x + \beta_i y + \gamma_i z) ds = S_1 + S_2 + S_3 = \\ &\iint_{D_i} A \sqrt{1 + (\frac{\partial z}{\partial x})^2 + (\frac{\partial z}{\partial y})^2} dx dy \\ &+ \iint_{D_i} A \sqrt{1 + (\frac{\partial y}{\partial x})^2 + (\frac{\partial y}{\partial z})^2} dx dz \\ &+ \iint_{D_i} A \sqrt{1 + (\frac{\partial x}{\partial y})^2 + (\frac{\partial x}{\partial z})^2} dy dz \end{aligned} \tag{7.7}$$

where $A = (\alpha_i x + \beta_i y + \gamma_i z)$.

Here D_i is the projected area of Δ_i in the coordinate plane. We choose the first area integral in (7.7), to explain the calculation. The equation of the triangular surface is $a_i x + b_i y + c_i z - d = 0$. If $a_i = b_i = 0$, then $z = \dfrac{d}{c_i}$. The triangular mesh is parallel to the X–Y plane. The projected area D_i of the triangular mesh S_i is shown in Fig. 7.4. In such a case, the area integral can be given as

$$\begin{aligned} S_1 &= \iint_{D_i} (\alpha_i x + \beta_i y + \gamma_i z) \sqrt{1 + (\frac{\partial z}{\partial x})^2 + (\frac{\partial z}{\partial y})^2} dx dy \\ &= \iint_{D_1} z dx dy = \iint_{D_1} \frac{d}{c_i} dx dy \end{aligned} \tag{7.8}$$

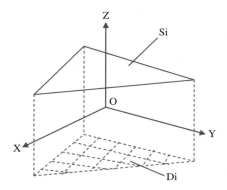

Fig. 7.4. Projection of triangular mesh S_i

If a_i or $b_i \neq 0$, the triangular mesh surface integral is the sum of the three projected area integrals. The other area integrals are calculated in a similar way to the first and will not be detailed here.

After all the triangular mesh surface integrals are calculated, if newly obtained data points do not bring obvious changes to the previous measurement value of the surface integral, i.e.

$$\left| V_i - V_{i+1} \right| = \left| \iiint_{Vi} dV - \iiint_{Vi+1} dV \right| < \Delta \tag{7.9}$$

then the process of measurement will be terminated.

Here V_i, V_{i+1} are the volume values at two successive viewpoints i and i+1 respectively. V_{i+1} in (7.9) is the data points encompassed volume at the (i + 1)th viewpoint, and is given as

$$V_{i+1} = V_i + V'_{i+1} - \sum_{j=1}^{i-1} V_{D_i \cap D_{i+1}} \tag{7.10}$$

where $V_i = \sum_{j=1}^{i} V_j$ is the data points encompassed volume from the previous viewpoint i. V'_{i+1} is the encompassed volume from data points acquired at the (i + 1)th viewpoint. $\sum_{j=1}^{i-1} V_{D_j \cap D_{j+1}}$ is the encompassed volume from data points of the overlapped areas at the (i + 1)th viewpoint and the previous viewpoints i. Each time a new triangle is generated, we will check for its possible intersection with the neighboring meshes. If it is possible, the current re-triangulation will be stopped and instead a vertex of the intersected edge will be chosen to generate another new triangle. This will lead to a change in the volume integral in the overlapped areas.

Equation (7.9) can be rewritten as

$$\left| V_{i+1} - V_i \right| = \left| V_{i+1}' - \sum_{j=1}^{i-1} V_{D_j \cap D_{j+1}} \right| = \Delta \tag{7.11}$$

where Δ is the variation in volume between the $(i + 1)$th viewpoint and the re-integral volume value from the overlapped surface data at the $(i + 1)$th viewpoint and the preceding viewpoints i.

Following (7.11), the termination condition is based on the difference between the surface integral of the newly acquired surface data and the overlapped area surface integral. In practical implementation, if there are no overlapping areas between viewpoints during the view planning, we skip the calculation of the surface integral. Otherwise, we only calculate the surface integral for new data obtained from the new viewpoint and the overlapped areas. With more views taken, the overlapped areas will increase, resulting in smaller and smaller variations in the volume (approaching zero if there are no calculation errors). In practice, there are always some errors including the matching errors, quantization errors and system calibration errors. All these errors would lead to some none-zero residue in the solution of (7.11). However, if the surface data acquired from a new viewpoint totally overlap those from the previous viewpoints, the above error residue will be very small, i.e.

$$\left| V_{i+1} - V_i \right| = \left| \iiint_{Vi+1} dV - \iiint_{Vi} dV \right| \approx \Delta < \delta$$

When a set error tolerance is satisfied, the measurement and view planning process will be terminated.

7.4 Experiments

The view planning system was implemented as part of an automated surface acquisition system. Several experiments were carried out in our laboratory on the construction of object models. The acquisition process contains the following steps: scan, register the new data with the partial model available, integrate the new data with the partial model, judge if the measurement is to be continued, determine the NBV, and repeat the above until the termination condition is satisfied.

The range sensing is achieved by using a camera and projector to form an active vision system (Li and Liu 2003). The sensor can move around the object. As the object is placed on a worktable, the movement of the sensor in space is constrained to the upper semi-sphere space. The system is illustrated in Fig. 7.5. A head model (shown in Fig. 7.6) was used as a typical example to illustrate the implementation of the developed view planning method for model construction. The first view was

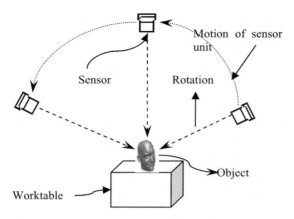

Fig. 7.5. A view of the scan setup

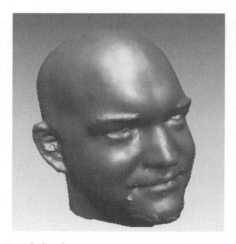

Fig. 7.6. The test object: male head

assumed to be taken from an arbitrary viewpoint. The succeeding next views for the unseen part of the object were determined by the trend surface method (which will be further discussed in detail in Chap. 9).

The object model was incrementally built by five views. At each view, a new surface part was acquired and integrated with the previously acquired data to form a partial model. The exploration direction and sensor pose were determined by the trend surface method. The planned views and the part of the surface acquired at each view is shown in Fig. 7.7 (a–e). The viewpoint vector has a format of $[x, y, z, \alpha, \beta, \varphi]$, representing the six parameters of the 3D position and 3-axis rotation.

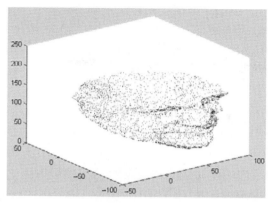

(**a**) Viewpoint 1 (−329.39, −534.14, 24.268, −1.0182, −0.55259, 0.038653)

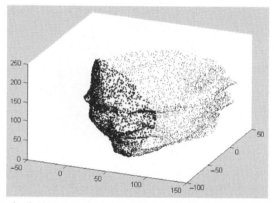

(**b**) Viewpoint 2 (414.4, −58.619, 4, 133.24, −1.4302, −0.14052, 0.22356)

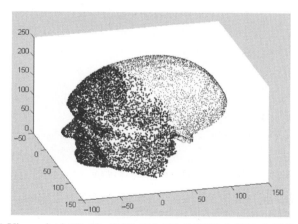

(**c**) Viewpoint 3 (51.48, 43.19, 615.489, 0.8727, 0.6981, 0.994)

Fig. 7.7. The planned views for a head model

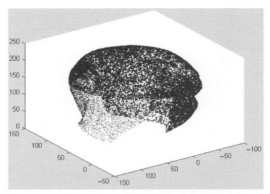

(**d**) Viewpoint 4 (−58.219, 373.31, 280.85, −1.4161, 0.1105, 0.4889)

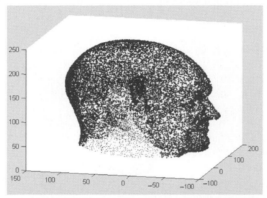

(**e**) Viewpoint 5 (−563.05, −399.39, 190.25, −0.953, −0.7093, 0.3249)

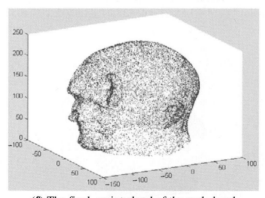

(**f**) The final   point cloud of the male head

Fig. 7.7. (Continued)

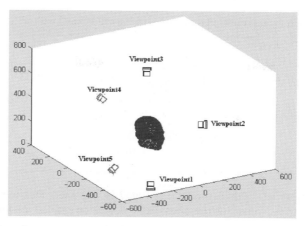

Fig. 7.7. (Continued)

In Fig. 7.7 (b–e), light blue represents the new data acquired from the planned next view. Dark blue represents the available data acquired from the previous view(s). At the same time, the self-termination judgment method was also tested in viewpoint planning. The results in the surface integral and computation time are given in Table 7.1 for each viewpoint. Here "#Views" refers to the number of viewpoints required to obtain the complete object surface. "Max time" is the time it takes for the system to calculate the surface integrals. Figure 7.8 shows the variations in the volumes between two successive viewpoints. Here, the sixth viewpoint is used only to show that the self-termination judgment method is convergent. This last viewpoint is not needed in a real implementation. As can be seen, the final variation in volume approaches approximately zero, indicating that the termination condition is satisfied.

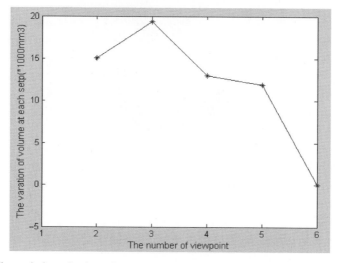

Fig. 7.8. The variation of volume between two successive viewpoints

Table 7.1. Computation results in reconstructing the head model with self-termination

View steps	1	2	3	4	5	6
CPU time(s)	16.621	46.102	78.23	98.34	113.75	128.34
Surface integral (104mm3)	2.2516	3.7583	5.86954	6.9974	8.1869	8.1898
Volume change (104mm3)		1.5067	1.9371	1.3020	1.1895	0.0029

In yet another test, a duck model (see Fig. 7.9) was used. Due to the complexity of this object, in some surface areas, the occlusion-guided method is used to continue the view planning and acquisition. The results are shown in Fig. 7.10.

The duck model was incrementally built by six views. During the view planning process, the sensor pose for Viewpoint2 could not be determined by the trend surface method. In Fig. 7.10a, the dark color represents the real surface data, and the dark blue represents the acquired data from viewpoint one. The light blue represents the trend surface. As can be seen, the prediction by the trend surface is invalid for the next view planning. This was identified by our algorithm. As a result, the occlusion-guided method was chosen automatically by the system to continue the view planning.

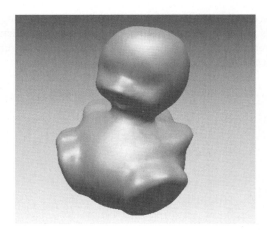

Fig. 7.9. The test object: Duck model

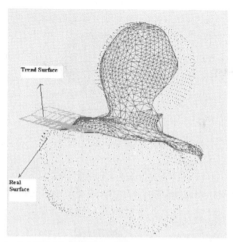

(**a**) Viewpoint1 (45.665, 405.41, 139.21, 0.079731, 89.92, −0.23819)

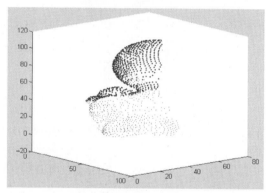

(**b**) Viewpoint2 (481.27, −417.07, 14.332, 0.95439, 89.046, 0.024294)

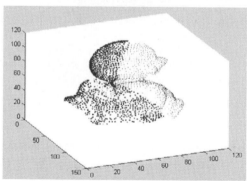

(**c**) Viewpoint3 (93.497, 316.81, 383.88, 0.21023, 89.79, 0.70843)

Fig. 7.10. The acquisition procedure of a duck model

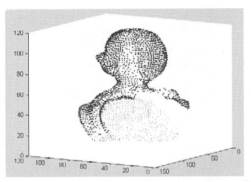

(**d**) Viewpoint4 (−58.186, 405.47, 138.89, 0.10165, 89.898, 0.23764)

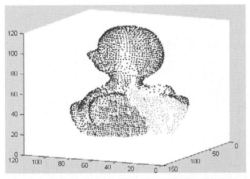

(**e**) Viewpoint5 (−41.05, 61.126, 583.63, 0.49481, 89.505, 1.4238)

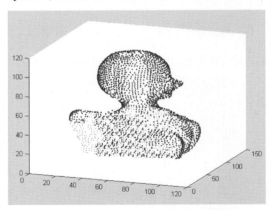

(**f**) Viewpoint6 (−246.49, 348.26, 324.86, 0.52414, 89.476, 0.5831)

Fig. 7.10. (Continued)

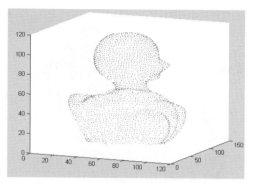

(g) The final point cloud of the duck model

Fig. 7.10. (Continued)

The self-termination judgment method in viewpoint planning is tested on the duck model, with the result in volume changes given in Fig. 7.11. As observed in this case, although the trend surface method can be invalid for complex objects, the invalid areas can be identified by our method and then another method can be used to complete the next best view planning. In all the test cases, the view planning was successfully completed without knowing the object model and complete object models were automatically acquired without any human interference.

It should be noted that in the above experiments, we assumed that no model of the object was available for the view planning. If we have the model of the target, careful planning will result in a minimum number of viewpoints via some optimization approaches, as each view including the first can be planned optimally. Take the head model for example, if we have the model, four views will enable us to acquire its surface, as shown in Fig. 7.12. Without an object model, however, the

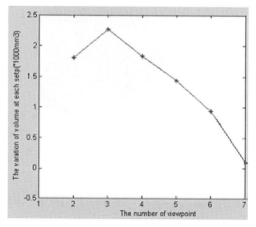

Fig. 7.11. The variation of volume between two successive viewpoints in the duck model

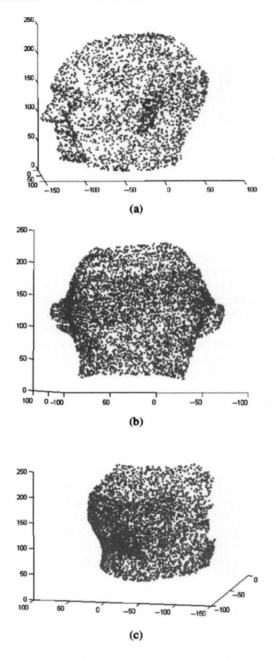

Fig. 7.12. Acquisition of a known model

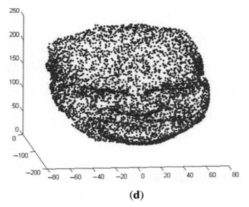

(d)

Fig. 7.12. (Continued)

first view is arbitrary in space and the information thus acquired is normally less than the former case. Therefore, when the target is unknown, more views will generally be needed than the case when it is known.

To compare our method with a previous one, we implemented Maver's algorithm [1993] for the duck model. Figure 7.13 shows some results. Here, it turned out that the first and second views were the same as our algorithm's (i.e. Fig. 7.13 (a),(b) are the same as Fig. 7.12 (a),(b)). At each viewpoint, there are a number of occlusion polygons, resulting in a number of new viewing directions. In this case, Maver's algorithm yields a position from which the detected occluded area is maximal. The next viewpoints were estimated based on the occlusion information extracted from the previous images.

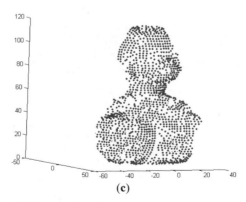

(c)

Fig. 7.13. Acquisition with Maver's algorithm

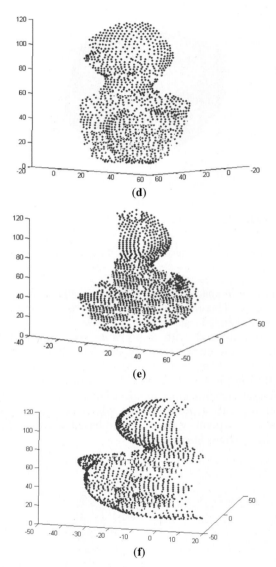

Fig. 7.13. (Continued)

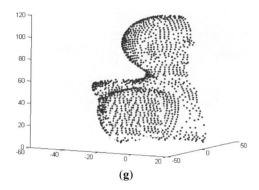

(g)

Fig. 7.13. (Continued)

Table 7.2. Computation results in reconstructing the duck model with Maver's algorithm

View steps	3	4	5	6	7
CPU time(s)	93	132	187	213	265
Surface integral ($10^3 mm^3$)	2.398	3.761	4.97	6.533	8.185
Integral change ($10^3 mm^3$)	0.895	1.363	1.209	1.563	1.652

Although Maver's algorithm could be used to obtain a good model automatically, the number of needed viewpoints is larger than our algorithm's (Table 7.2). This is because the viewpoints are acquired based on the occlusion information in Maver's algorithm. The object surface information (the surface continuity or surface order) is not used in formulating the next view. Such information could improve the view planning.

It should also be noted that our view planning method takes into account the surface trend without excessively considering the surface details. For example, the area of the ear in the head model has some complex and concave surface features (see the Fig. 7.14). This leads to some occlusions at certain viewpoints so that some features could be lost. However, these features are small and will not make contributions large enough to affect the view planning. In the example about the area of the ear, the un-scanned surface is normally less than 1% of the whole area seen in that particular view. On the other hand, if these features are big enough, e.g. with serious occlusions or large areas of abrupt changes in surface geometry as in the duck model, they will affect the algorithm via invoking an alternative occlusion-based method. Concave surfaces in general still present a challenging task for view planning with issues open for future research.

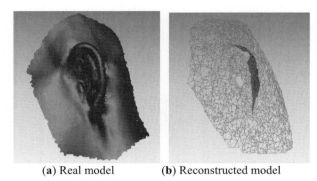

(a) Real model   (b) Reconstructed model

Fig. 7.14. Comparison in local surface features

7.5 Summary

For a system to be able to automatically reconstruct a 3D model, it not only needs to be able to plan the sensor viewpoints but also to terminate the reconstruction by itself during the model acquisition process. In this chapter, an approach to self-termination for measurement and digitalization of 3D objects of arbitrary shapes is presented. With this method, all measurement data acquired at each viewpoint are used in checking the termination condition. By calculating the volume encompassed by data points acquired at each view and analyzing the variation of the volume at two successive viewpoints, this method provides reliable judgment on whether the termination criterion is satisfied. During the process of measurement, as the surface data are not complete to form a closed shape, it is difficult to directly calculate the volume encompassed by the surface data. Nevertheless, the volume of a data cloud could be achieved via computing the surface integral of the triangular meshes. In this way, a measurement task will continue until the volume computed in this way no longer changes at a new viewpoint.

The self-termination judgment method is based on Gauss' Theorem by checking the variations of the surface integrals between two successive viewpoints. When the variation is smaller than a given threshold, the view planning and data acquisition process will be terminated. Based on this principle, the termination occurs only when no new surface data are acquired at a new viewpoint. With this method, all surface data acquired up to the current viewpoint are used to check the termination condition. This overcomes the limitations of previous methods in using only part of the available surface data in the termination condition. As a result, the proposed method is robust to shape variations of the object surface. The implementation results show that the proposed method is effective in automatic reconstruction of the surface models of unknown objects.

This chapter has mentioned the method of trend surface, which is a strategy for generating a sequence of viewing poses. The method involves decisions on the exploration direction and sensor poses by using trend surface as the cue to predict the unknown portion of an object surface. With the exploration direction determination, the unknown surface of an object is predicted by the trend surface. Then the pose of the next viewpoint is obtained by imposing the sensor placement constrains. More details regarding this technology are going to be presented in Chaps. 8 and 9 in detail.

Chapter 8
Information Entropy Based Planning

In this chapter, we present an approach with information entropy based sensor planning for reconstruction of freeform surfaces of 3D objects. To achieve the reconstruction, the object is first sliced into a series of cross-section curves, with each curve to be reconstructed by a closed B-spline curve. In the framework of Bayesian statistics, we propose an improved Bayesian information criterion (BIC) for determining the B-spline model complexity. Then, we analyze the uncertainty of the model using entropy as the measurement. Based on this analysis, we predict the information gain for each cross section curve for the next measurement. After predicting the information gain of each curve, we obtain the information change for all the B-spline models. This information gain is then mapped into the view space. The viewpoint that contains maximal information gain about the object is selected as the Next Best View. Experimental results show successful implementation of the proposed view planning method for digitization and reconstruction of freeform objects.

8.1 Overview

This chapter presents an information entropy based viewpoint planning method for the digitization and reconstruction of a 3D freeform object. The object is sliced into a set of cross section curves and a closed B-spline curve is used to reconstruct each cross section curve by fitting to partial data points. An information criterion is developed for selecting the B-spline model structure. Based on the selected B-spline model, we use information entropy as the uncertainty measure of the B-spline model and analyze the uncertainty of each B-spline cross section curve to predict the information gain for new measurements to be taken. As a result, we can obtain the prediction of the information gain about the object. The information gain is then mapped to the view space. The view that has the maximal information gain on the object is then selected as the Next Best View (NBV). The proposed information entropy based viewpoint planning procedure is illustrated in Fig. 8.1.

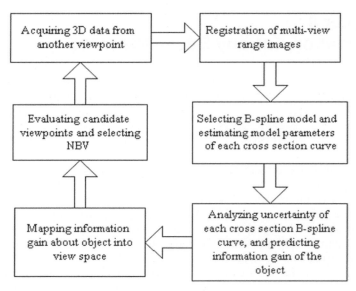

Fig. 8.1. Information entropy based viewpoint planning

This work is novel concerning the parameter estimation for the NBV problem. In contrast to Whaite's method (1997), here we analyze and reconstruct a B-spline model in the framework of Bayesian statistics. The B-spline model is more powerful in describing objects than a super-ellipsoid. In addition, we introduce the principle of model selection by which the proposed improved BIC criterion makes the B-spline model adaptable when newly acquired data are available. The rest of this chapter is organized as follows. In Sect. 8.2, we describe the reconstruction of cross section curves with closed B-splines and introduce the modified BIC for selecting a B-spline model structure. In Sect. 8.3, we define the information entropy of B-spline model to analyze its uncertainty and predict the information gain on an object. In Sect. 8.4, we evaluate the visibility of candidate viewpoints for selecting NBV. Finally, we present the experimental results in implementing the proposed method in Sect. 8.5 followed by conclusions in Sect. 8.6.

8.2 Model Representation

For object surface reconstructions, the 3D shape can be divided into a series of cross section curves each representing the local geometrical feature of the object. These cross section curves can be described by a set of parametric equations. For reconstruction purposes using parametric equations, the most common methods include spline functions (e.g. B-splines) (Fernand and Wang 1994), implicit polynomials and superquadrics (e.g. superellipsoids) (Whaite 1997). Compared with implicit polynomials and superquadrics, B-splines have the following main advantages:

- *Smoothness and continuity*, which allows a curve to consist of a concatenation of curve segments, yet be treated as a single unit;
- *Built-in boundedness*, a property which is lacking in implicit or explicit polynomial representation whose zero set can shoot to infinity;
- *Parameterized representation*, which decouples the x, y coordinates to be treated separately.

8.2.1 Curve Approximation

Let a closed cubic B-spline curve consist of $n+1$ curve segments, defined by

$$\mathbf{p}(t) = \sum_{j=0}^{n+3} B_{j,4}(t) \cdot \mathbf{\Phi}_j, \tag{8.1}$$

where $\mathbf{p}(t) = [x(t), y(t)]$ is a point on the B-spline curve with location parameter t. $B_{j,4}(t)$ is the jth normalized cubic B-spline basis function defined over the following uniform knots vector

$$[u_{-3}, u_{-2}, u_{-1}, u_0, \ldots, \ldots, u_{n+4}] = [-3, -2, -1, 0, \ldots, n+4]. \tag{8.2}$$

The amplitude of $B_{j,4}(t)$ is in the range of $(0.0, 1.0)$, and the support region of $B_{j,4}(t)$ is compact and nonzero for $t \in [u_j, u_{j+4}]$. $(\mathbf{\Phi}_j)_{j=0}^{n+3}$ are the cyclical control points which satisfy the following conditions

$$\mathbf{\Phi}_{n+1} = \mathbf{\Phi}_0, \ \mathbf{\Phi}_{n+2} = \mathbf{\Phi}_1, \ \mathbf{\Phi}_{n+3} = \mathbf{\Phi}_2 \tag{8.3}$$

For a set of m data points $\mathbf{r} = (\mathbf{r}_i)_{i=1}^m = ([x_i, y_i])_{i=1}^m$, let d^2 be the sum of the squared residual errors between the data points and their corresponding points on the B-spline curve, i.e.

$$d^2 = \sum_{i=1}^m \|\mathbf{r}_i - \mathbf{p}(t_i)\|^2 = \sum_{i=1}^m \left[\mathbf{r}_i - \sum_{j=0}^{n+3} \mathbf{B}_{j,4}(t_i) \cdot \mathbf{\Phi}_j\right]^2. \tag{8.4}$$

From the cyclical condition of control points in (8.3), there are only $n+1$ control points to be estimated. The LS estimation of the $n+1$ control points are obtained from the curve points by minimizing d^2 in (8.4) with respect to $\mathbf{\Phi} = [\mathbf{\Phi}_x^T, \mathbf{\Phi}_y^T]^T = [\mathbf{\Phi}_{x0}, \ldots \mathbf{\Phi}_{xn}, \mathbf{\Phi}_{y0}, \ldots, \mathbf{\Phi}_{yn}]^T$. By factorization of the B-spline, two separate solutions are obtained in the matrix as follows

$$\begin{cases} \boldsymbol{\Phi}_x = [\mathbf{B}^T \mathbf{B}]^{-1} \mathbf{B}^T \mathbf{x} \\ \boldsymbol{\Phi}_y = [\mathbf{B}^T \mathbf{B}]^{-1} \mathbf{B}^T y \end{cases} \tag{8.5}$$

where $\mathbf{x} = [x_1, \ldots, x_m]^T$, $\mathbf{y} = [y_1, \ldots, y_m]^T$,

$$\mathbf{B} = \begin{bmatrix} \overline{B}_{0,4}^1 + \overline{B}_{n+1,4}^1 & \overline{B}_{1,4}^1 + \overline{B}_{n+2,4}^1 & \overline{B}_{2,4}^1 + \overline{B}_{n+3,4}^1 & \cdots & \overline{B}_{n,4}^1 \\ \overline{B}_{0,4}^2 + \overline{B}_{n+1,4}^2 & \overline{B}_{1,4}^2 + \overline{B}_{n+2,4}^2 & \overline{B}_{2,4}^2 + \overline{B}_{n+3,4}^2 & \cdots & \overline{B}_{n,4}^2 \\ \vdots & \vdots & \vdots & \vdots & \vdots \\ \overline{B}_{0,4}^m + \overline{B}_{n+1,4}^m & \overline{B}_{1,4}^m + \overline{B}_{n+2,4}^m & \overline{B}_{2,4}^m + \overline{B}_{n+3,4}^m & \cdots & \overline{B}_{n,4}^m \end{bmatrix},$$

and $\overline{B}_{j,4}^i = B_{j,4}(t_i)$.

Here, we adopt the chord length method, which is the most popular one, for the parameterization of the B-spline. The chord length L of a curve is calculated as follows

$$L = \sum_{i=2}^{m+1} \left\| \mathbf{r}_i - \mathbf{r}_{i-1} \right\| \tag{8.6}$$

where $\mathbf{r}_{m+1} = \mathbf{r}_1$ for a closed curve. The t_i associated with the point q_i is given as

$$t_i = t_{i-1} + \frac{\left\| \mathbf{r}_i - \mathbf{r}_{i-1} \right\|}{L} \cdot t_{max} \tag{8.7}$$

where $t_1 = 0$ and $t_{max} = n + 1$.

8.2.2 Improved BIC Criterion

It is known that for a given set of measurement data, there exists a model of optimal complexity corresponding to the smallest prediction (generalization) error for further data. The complexity of a B-spline model of a surface is related to its control point (parameter) number (Fernand and Wang 1994). If the B-spline model is too complicated, the approximated B-spline surface tends to over-fit noisy measurement data. If the model is too simple, then it is not capable of fitting the measurement data, making the approximation results under-fitted. The problem of finding an appropriate model, referred to as model selection, is important for achieving a high level generalization capability. Model selection has been studied from various standpoints in the field of statistics, including information statistics, Bayesian statistics, and structural risk minimization. The Bayesian approach (Djuric 1998, Torr 2002) is perhaps the most general and most powerful model selection method. Based on posterior model probabilities, the Bayesian approach estimates a probability distribution over an ensemble of models. The prediction is accomplished by averaging over the ensemble of models. Accordingly, the uncertainty of the models is taken into account, and complex models with more degrees of freedom are penalized.

Given a set of models $\{M_k, k = 1, 2, \ldots, k_{max}\}$ and data $\mathbf{r}$, the Bayesian approach selects the model with the largest posterior probability. The posterior probability of model M_k is

$$p(M_k \mid \mathbf{r}) = \frac{p(\mathbf{r} \mid M_k)p(M_k)}{\sum_{L=1}^{k_{max}} p(\mathbf{r} \mid M_L)p(M_L)} \tag{8.8}$$

where $p(\mathbf{r} \mid M_k)$ is the integrated likelihood of model M_k and $p(M_k)$ is the prior probability of model M_k. To find the model with the largest posterior probability, we evaluate $p(M_k \mid \mathbf{r})$ for $k = 1, 2, \ldots, k_{max}$ and select the model that has the maximum $p(M_k \mid \mathbf{r})$, that is

$$M = \arg \max_{M_k, k=1,\ldots,k_{max}} \{p(M_k \mid \mathbf{r})\}$$

$$= \arg \max_{M_k, k=1,\ldots,k_{max}} \left\{ \frac{p(\mathbf{r} \mid M_k)p(M_k)}{\sum_{L=1}^{k_{max}} p(\mathbf{r} \mid M_L)p(M_L)} \right\}. \tag{8.9}$$

Here, we assume that the models have the same likelyhood a priori, so that $p(M_k) = 1/k_{max}$, $(k = 1, 2, \ldots, k_{max})$. Therefore, the model selection in (8.8) will not be affected by $p(M_k)$. This is also the case with $\sum_{L=1}^{k_{max}} p(r \mid M_L)p(M_L)$ since it is not a function of M_k. Consequently, we can ignore the factors $p(M_k)$ and $\sum_{L=1}^{k_{max}} p(r \mid M_L)p(M_L)$ in computing the model criteria. Equation (8.9) then becomes

$$M = \arg \max_{M_k, k=1,\ldots,k_{max}} \{p(\mathbf{r} \mid M_k)\}. \tag{8.10}$$

To calculate the posterior probability of model M_k, we need to evaluate the marginal density of data for each model $p(\mathbf{r} \mid M_k)$, which requires multidimensional integration

$$p(\mathbf{r} \mid M_k) = \int_{\boldsymbol{\Phi}_k} p(\mathbf{r} \mid \boldsymbol{\Phi}_k, M_k)p(\boldsymbol{\Phi}_k \mid M_k)d\boldsymbol{\Phi}_k \tag{8.11}$$

where $\boldsymbol{\Phi}_k$ is the parameter vector for model M_k, $p(\mathbf{r} \mid \boldsymbol{\Phi}_k, M_k)$ is the likelihood and $p(\boldsymbol{\Phi}_k \mid M_k)$ is the prior distribution for model M_k.

In practice, calculating the multidimensional integration is very hard, especially for obtaining a closed-form analytical solution. The research in this area has resulted in many approximation methods for achieving this. The Laplace's approximation method for the integration appears to be a simple one and has become a standard method for calculating the integration of multi-variable Gaussians (Torr 2002). This yields

$$
\begin{aligned}
p(\mathbf{r} \mid M_k) &= \int_{\Phi_k} p(\mathbf{r} \mid \Phi_k, M_k) p(\Phi_k \mid M_k) d\Phi_k \\
&\cong (2\pi)^{d_k/2} \mid H(\hat{\Phi}_k) \mid^{-1/2} p(\mathbf{r} \mid \hat{\Phi}_k, M_k) p(\hat{\Phi}_k \mid M_k)
\end{aligned}
\tag{8.12}
$$

where $\hat{\Phi}_k$ is the maximum likelihood estimate of Φ_k, d_k denotes the number of parameters (control points for B-spline model) in model M_k, and $H(\hat{\Phi}_k)$ is the Hessian matrix of $-\log p(\mathbf{r} \mid \Phi_k, M_k)$ evaluated at $\hat{\Phi}_k$,

$$
H(\hat{\Phi}_k) = -\frac{\partial^2 \log p(\mathbf{r} \mid \Phi_k, M_k)}{\partial \Phi_k \partial \Phi_k^T} \Bigg|_{\Phi_k = \hat{\Phi}_k}
\tag{8.13}
$$

This approximation is particularly good when the likelihood function is highly peaked around $\hat{\Phi}_k$. This is usually the case when the number of data samples is large. Neglecting the terms of $p(\hat{\Phi}_k \mid M_k)$ and using log in the calculation, the posterior probability of model M_k becomes

$$
M = \arg \max_{M_k, k=1,\ldots,k_{\max}} \left\{ \log p(\mathbf{r} \mid \hat{\Phi}_k, M_k) - \frac{1}{2} \log \mid H(\hat{\Phi}_k) \mid \right\}
\tag{8.14}
$$

The likelihood function $p(\mathbf{r} \mid \hat{\Phi}_k, M_k)$ of a closed B-spline cross section curve can be factored into x and y components as

$$
p(\mathbf{r} \mid \hat{\Phi}_k, M_k) = p(\mathbf{x} \mid \hat{\Phi}_{kx}, M_k) \cdot p(\mathbf{y} \mid \hat{\Phi}_{ky}, M_k)
\tag{8.15}
$$

where $\hat{\Phi}_{kx}$ and $\hat{\Phi}_{ky}$ can be calculated by (8.5).

Consider the x component. Assuming that the residual error sequence is zero mean and white Gaussian with variance $\sigma_{kx}^2(\hat{\Phi}_{kx})$, we have the following likelihood function

$$
\begin{aligned}
p(\mathbf{x} \mid \hat{\Phi}_{kx}, M_k) &= \left(\frac{1}{2\pi\sigma_{kx}^2(\hat{\Phi}_{kx})} \right)^{m/2} \\
&\quad \exp\left\{ -\frac{1}{2\sigma_{kx}^2(\hat{\Phi}_{kx})} \sum_{k=0}^{m-1} [x_k - \mathbf{B}_k \hat{\Phi}_{kx}]^2 \right\}
\end{aligned}
\tag{8.16}
$$

with $\sigma_{kx}^2(\hat{\boldsymbol{\Phi}}_{kx}, M_k)$ estimated by

$$\hat{\sigma}_{kx}^2(\hat{\boldsymbol{\Phi}}_{kx}) = \frac{1}{m}\sum_{k=0}^{m-1}[x_k - \mathbf{B}_k\hat{\boldsymbol{\Phi}}_{kx}]^2 \qquad (8.17)$$

Similarly, the likelihood function of the y component can also be obtained. The corresponding Hessian matrix $\hat{H}_k$ of $-\log p(\mathbf{r} \mid \boldsymbol{\Phi}_k, M_k)$ evaluated at $\hat{\boldsymbol{\Phi}}_k$ is

$$H(\hat{\boldsymbol{\Phi}}_k) = \begin{bmatrix} \dfrac{\mathbf{B}^T\mathbf{B}}{\hat{\sigma}_{kx}^2(\hat{\boldsymbol{\Phi}}_{kx})} & \mathbf{0} \\ \mathbf{0} & \dfrac{\mathbf{B}^T\mathbf{B}}{\hat{\sigma}_{ky}^2(\hat{\boldsymbol{\Phi}}_{ky})} \end{bmatrix} \qquad (8.18)$$

Approximating $\dfrac{1}{2}\log|H(\hat{\boldsymbol{\Phi}}_k)|$ by the asymptotic expected value of Hessian $\dfrac{1}{2}(d_{kx} + d_{ky})\log(m)$, we can obtain the Bayesian information criterion (BIC) for selecting the structure of the B-spline curve

$$M = \arg\max_{M_k, k=1,\ldots,k_{\max}} \left\{ \begin{matrix} -\dfrac{m}{2}\log\hat{\sigma}_{kx}^2(\hat{\boldsymbol{\Phi}}_{kx}) - \dfrac{m}{2}\log\hat{\sigma}_{ky}^2(\hat{\boldsymbol{\Phi}}_{ky}) \\ -\dfrac{1}{2}(d_{kx} + d_{ky})\log(m) \end{matrix} \right\} \qquad (8.19)$$

where d_{kx} and d_{ky} are the number of control points in the x and y directions respectively, m is the number of data points.

In the conventional BIC criterion as shown in (8.19), the first two terms measure the estimation accuracy of the B-spline model. In general, the variance $\hat{\sigma}_k^2$ estimated from (8.17), tends to decrease with the increase in the number of control points. The smaller the variance value in $\hat{\sigma}_k^2$, the bigger the value of the first two terms (as the variance is much smaller than one) and therefore the higher the order (i.e. the more control points) of the model resulting from (8.19). However, if too many control points are used, the B-spline model will over-fit noisy data points. An over-fitted B-spline model will have a poor generalization capability. Model selection thus should achieve a proper tradeoff between the approximation accuracy and the number of control points of the B-spline model. With a conventional BIC criterion, the same data set is used for estimating both the control points of the B-spline model and the variances. Thus the first two terms in (8.19) cannot detect the occurrence of over fitting in the B-spline model selected. In theory, the third term in (8.19) could penalize over-fitting as it appears directly proportional to the number of control points used. In practice, however, we note from our experience that the effect of this penalty term is insignificant compared with that of the first two terms. As a result, the conventional BIC criterion is rather insensitive to the occurrence of over-fitting and tends to select more control points in the B-spline model to approximate the data point, which normally results in a model with poor generalization capability.

The reason for the occurrence of over-fitting in a conventional BIC criterion lies in the way the variances σ_{kx}^2 and σ_{ky}^2 are obtained. A reliable estimate of σ_{kx}^2 and σ_{ky}^2 should be based on re-sampling of the data. In other words, the generalization capability of a B-spline model should be validated using another set of data points rather than the same data used in obtaining the model. To achieve this, we divide the available data into two sets: a training sample and a prediction sample. The training sample is used only for model estimation, whereas the prediction sample is used only for estimating the data noise σ_{kx}^2 and σ_{ky}^2. For a candidate B-spline model M_k with d_{kx} and d_{ky} control points in x and y directions, the BIC in (8.19) is thus evaluated via the following two steps:

1. Estimate the model parameter $\hat{\mathbf{\Phi}}_k$ using the training sample by (8.5);

2. Estimate the data noise σ_k^2 using the prediction sample by (8.17).

If the model $\hat{\mathbf{\Phi}}_k$ fitted to the training data is valid, then the estimated variance $\hat{\sigma}_k^2$ from the prediction sample should also be a valid estimate of the data noise. If the variance $\hat{\sigma}_k^2$ found from the prediction sample is unexpectedly large, we have reasons to believe that the candidate model fits the data badly. It can be seen that the data noise $\hat{\sigma}_k^2$ estimated from the prediction sample will thus be more sensitive to the quality of the model than the one directly estimated from the training sample, as the variance σ_k^2 estimated from the prediction sample also has the capability of detecting the occurrence of over-fitting.

8.3 Expected Error

In Sect. 8.2, we described our approach to model selection and parameter estimation in the framework of Bayesian statistics. In this section, we will discuss how the same framework for B-spline curve approximation relates to the task of selecting the NBV for acquiring new data. For simplification of the description, we will replace $\mathbf{\Phi}_k$ by $\mathbf{\Phi}$ to show that we are dealing with the selected "best" B-spline model with d_{kx} and d_{ky} control points. To obtain the approximate B-spline model, we will predict the distribution of the information gain on the model's parameter $\mathbf{\Phi}$ along each cross section curve. A measure of the information gain will be designed whose expected value will be maximal when the new measurement data are acquired. The measurement is based on Shannon's entropy whose properties make it a sensible information measure here. We will describe the information entropy of the B-spline model and how to use it to achieve maximal information gain on the parameters of the B-spline model $\mathbf{\Phi}$.

8.3.1 Information Entropy of a B-Spline Model

Given Φ and the data points $\mathbf{r} = (\mathbf{r}_i)_{i=1}^m$ which are assumed to be statistically independent, with Gaussian noise of zero mean and variance σ^2, the joint probability of $\mathbf{r} = (\mathbf{r}_i)_{i=1}^m$ is

$$p(\mathbf{r} \mid \Phi) = \frac{1}{(2\pi\sigma^2)^{m/2}} \cdot \exp[-\frac{1}{2\sigma^2}(\mathbf{r} - \mathbf{B} \cdot \Phi)^T(\mathbf{r} - \mathbf{B} \cdot \Phi)] \qquad (8.20)$$

Equation (8.20) has an asymptotic approximation representation defined by Subrahmonia et al. (1996)

$$p(\mathbf{r} \mid \Phi) \approx p(\mathbf{r} \mid \hat{\Phi}) \exp[-\frac{1}{2}(\Phi - \hat{\Phi})^T H_m(\Phi - \hat{\Phi})] \qquad (8.21)$$

where $\hat{\Phi}_k$ is the maximum likelihood estimation of Φ given the data points and H_m is the Hessian matrix of $-\log p(\mathbf{r} \mid \Phi)$ evaluated at $\hat{\Phi}$ given data points $\mathbf{r} = (\mathbf{r}_i)_{i=1}^m$.

The posteriori distribution $p(\Phi \mid \mathbf{r})$ of the given data is approximately proportional to

$$p(\Phi \mid \mathbf{r}) \approx p(\mathbf{r} \mid \hat{\Phi}) \cdot \exp[-\frac{1}{2}(\Phi - \hat{\Phi})^T H_m(\Phi - \hat{\Phi})]p(\Phi) \qquad (8.22)$$

where the $p(\Phi)$ is the priori probability of the B-spline model parameters. If the priori has a Guassian distribution with mean $\hat{\Phi}$ and covariance H_m^{-1}, we have

$$p(\Phi \mid \mathbf{r}) \propto \exp[-\frac{1}{2}(\Phi - \hat{\Phi})^T H_m(\Phi - \hat{\Phi})] \qquad (8.23)$$

From Shannon's information entropy, the conditional entropy of $p(\Phi \mid \mathbf{r})$ is defined by

$$E_m(\Phi) = \int p(\Phi \mid \mathbf{r}) \cdot \log p(\Phi \mid \mathbf{r}) d\Phi \qquad (8.24)$$

If $p(\Phi \mid \mathbf{r})$ obeys Guassian distribution, the corresponding entropy is Mackay (1991)

$$E_m = \Delta + \frac{1}{2}\log(\det H_m^{-1} \qquad (8.25)$$

where Δ is a constant.

The entropy in (8.25) measures the information about the B-spline model parameters, given data points $(\mathbf{r}_1,\ldots,\mathbf{r}_m)$. The more information about Φ, the smaller the entropy will be. In this work, we use the entropy in (8.25) as the measurement of the uncertainty of the model parameter Φ. Thus, to minimize E_m, we will make $\det(H_m^{-1})$ as small as possible.

8.3.2 Information Gain

In order to predict the distribution of the information gain, we assume a new data point $\mathbf{r}_{m+1}$ collected along a contour. The potential information gain is determined by incorporating the new data point $\mathbf{r}_{m+1}$. If we move the new point $\mathbf{r}_{m+1}$ along the contour, the distribution of the potential information gain along the whole contour can be obtained. Now, we will derive the relationship between the information gain and the new data point $\mathbf{r}_{m+1}$.

Assume that a new data point $\mathbf{r}_{m+1}$ has been collected. Let $P(\Phi \mid \mathbf{r}_1,\ldots,\mathbf{r}_m,\mathbf{r}_{m+1})$ be the probability distribution of model parameter Φ after a new point $\mathbf{r}_{m+1}$ is added. Its corresponding entropy is $E_{m+1}=\Delta+\frac{1}{2}\log(\det \hat{H}_{m+1}^{-1})$. The information gain then is

$$\Delta E = E_m - E_{m+1} = \frac{1}{2}\log\frac{\det H_m^{-1}}{\det H_{m+1}^{-1}} \tag{8.26}$$

From (8.18), the new data point $\mathbf{r}_{m+1}$ will incrementally update the Hessian matrix as follows

$$H_{m+1} \approx H_m + \begin{bmatrix} \dfrac{1}{\sigma_x^2}\cdot\overline{B}_{m+1}^T\overline{B}_{m+1} & \mathbf{0} \\ \mathbf{0} & \dfrac{1}{\sigma_y^2}\cdot\overline{B}_{m+1}^T\overline{B}_{m+1} \end{bmatrix} \tag{8.27}$$

where $\hat{\sigma}_{m+1}^2 \approx \hat{\sigma}_m^2$. $\overline{B}_{m+1}$ is defined by

$$\overline{B}_{m+1} = [\overline{B}_{0,4}^{m+1}+\overline{B}_{n+1,4}^{m+1}, \overline{B}_{1,4}^{m+1}+\overline{B}_{n+2,4}^{m+1},\ldots,\overline{B}_{n,4}^{m+1}]$$

The determinant of H_{m+1}

$$\det H_{m+1} \approx$$

$$\det[\mathbf{I} + \begin{bmatrix} \dfrac{1}{\hat{\sigma}_x^2} \cdot \overline{B}_{m+1}^T \overline{B}_{m+1} & \mathbf{0} \\ \mathbf{0} & \dfrac{1}{\hat{\sigma}_y^2} \cdot \overline{B}_{m+1}^T \overline{B}_{m+1} \end{bmatrix} H_m^{-1}] \cdot \det H_m$$

can be simplified to

$$\det H_{m+1} \approx (1 + \overline{B}_{m+1} \cdot [\mathbf{B}^T \mathbf{B}]^{-1} \cdot \overline{B}_{m+1}^T)^2 \cdot \det H_m \qquad (8.28)$$

Since $\det H^{-1} = 1/\det H$, (8.26) can be simplified to

$$\Delta E = \log(1 + \overline{B}_{m+1} \cdot [\mathbf{B}^T \mathbf{B}]^{-1} \cdot \overline{B}_{m+1}^T) \qquad (8.29)$$

Assuming that the new additional data point $\mathbf{r}_{m+1}$ travels along the contour, the resulting potential information gain of the B-spline model will change according to (8.29). In order to reduce the uncertainty of the model, we would like to have the new data point at such a location that the potential information gain attainable is largest. Therefore, after reconstructing the section curve by fitting partial data acquired from previous viewpoints, the Next Best Viewpoint should be selected as the one that senses those new data points which yield the largest possible potential information gain for the B-spline model.

8.4 View Planning

A view space is a set of 3D positions where the sensor (vision system) takes measurements. We assume that the 3D object is within the field of view and the depth of view of the vision system. The optical settings of the vision system are fixed. Based on these assumptions, the parameters of the vision system to be planned are the viewing pose of the sensor. In this section, the candidate viewpoints are represented in a spherical viewing space. The view space is usually a continuous spherical surface. To reduce the number of viewpoints used in practice, we discretize the surface by using the icosahedron method. In addition, we assume that the view space is centered around the object, and its radius is equal to an a priori specified distance from the sensor to the object. As shown in Fig. 8.2, since the optical axis of the sensor passes through the center of the object, the viewpoint can be represented by pan-tilt angles ϕ ([$-180°$, $180°$]) and θ ([$-90°$, $90°$]).

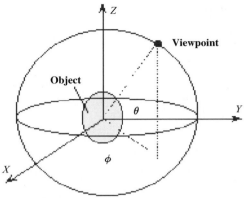

Fig. 8.2. Viewpoint representation

According to the representation of the viewing space, the fundamental task in the view planning here is to obtain the visibility regions in the viewing space that contain the candidate viewpoints where the missing information about the 3D object can be obtained without occlusions. The NBV should be the viewpoint that can give maximum information about the object.

With the above view space representation, we can now map the predicted information gain to the view space for viewpoint planning. For a viewpoint $v(\theta, \phi)$, we say one data point on the object is visible if the angle between its normal and the view direction is smaller than a breakdown angle α of the sensor. The view space V_k for each data point $\mathbf{r}_k$ ($k = 1, 2, \ldots$) is the set of all possible viewpoints that can see $\mathbf{r}_k$. The view space V_k can be calculated via the following procedure:

1. Calculating the normal vector $\mathbf{n}_k$ of a point $\mathbf{r}_k$ ($k = 1, 2, \ldots$) on the object, using a least square error fitting of a 3×3 local surface patch in its neighborhood.
2. Extracting viewpoints from which $\mathbf{r}_k$ is visible. These viewpoints are denoted as view space V_k.

After the view space V_k ($k = 1, 2, \ldots$) is extracted, we construct a measurement matrix **M.** The components $m_{k,j}$ of an l-by-w measurement matrix are given as

$$m_{k,j} = \begin{cases} \langle \mathbf{n}_k \cdot \mathbf{v}_j \rangle & \text{if } \mathbf{r}_k \text{ is visible to } v_j \\ 0 & \text{otherwise} \end{cases} \tag{8.30}$$

where $\mathbf{v}_j$ is the direction vector of viewpoint v_j.

Then, for each view $v(\theta, \phi)$, we define a global measure of the information gain $I(\theta, \phi)$ as the criterion to be summed over all visible surface points seen under this view of the sensor. $I(\theta, \phi)$ is defined by

$$I_j(\theta_j, \phi_j) = \sum_{k \in R_j} m_{k,j} \cdot \Delta E_k \tag{8.31}$$

where ΔE_k is the information gain at surface point $\mathbf{r}_k$, which is weighted by $m_{k,j}$.

Therefore, the Next Best View (θ^*, ϕ^*) is one that maximizes the information gain function of $I(\theta, \phi)$

$$(\theta^*, \phi^*) = \max_{\theta_j, \phi_j} I_j(\theta_j, \phi_j) \qquad (8.32)$$

8.5 Experiments

8.5.1 Setup

The information entropy based viewpoint planning algorithm is implemented as part of the work for 3D object reconstruction. The setup of a general 3D shape measurement system is schematically shown in Fig. 8.3. The sensor mounted on a robot consists of a projector that projects structured light onto the object and a CCD camera that captures the image of the illuminated object surface (Li and Liu 2003). This range sensor can give depth information of the scanned surface of an object in the form of a "data cloud". In the current implementation, the object is placed on a stationary platform. The robot has 6 DOF and is able to take a measurement of the object from any viewing pose specified within its work space. The modeling process for a 3D object consists of a sequence of four repeated steps: acquiring data on the object surface from a viewpoint, registering the acquired data, integrating the new data with a partial model and determining the NBV. This cycle will be repeated until the NBV terminates.

To slice the acquired "data cloud", we define an interval distance between cross section curves in a certain direction (e.g. the z direction) and project the data in the neighborhood of each cross section curve onto the plane on which the cross section curve lies. The preprocessing results of the 3D "data cloud" are shown in Fig. 8.4. Here the interval between two cross section curves was set at 1.5 mm and the neighborhood of each cross section curve is set at 0.2 mm.

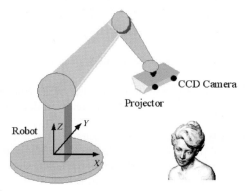

Fig. 8.3. System setup

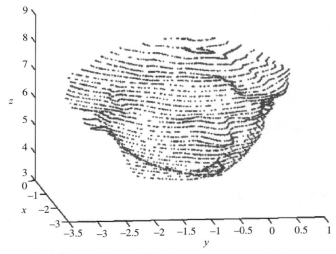

Fig. 8.4. Cross section curves after preprocessing

These points projected onto a cross section curve were distributed randomly. They need to be sorted out before the curve reconstruction can be performed. For each section curve, these projected points were transformed into the polar coordinate system. The phase angle was used to sort these points. To reconstruct these cross section curves via B-splines, we need to select an appropriate model structure first. The model selection is important for automated 3D modeling, to account for the data already acquired and to avoid over-fitting of the model.

8.5.2 Model Selection

In this section, the improved BIC criterion proposed will be used to select the B-spline model to represent the cross section curves. Two cross section curves from a series of sliced cross section curves will be used as examples to demonstrate the effectiveness of our approach. To evaluate the selected models, the following performance indexes are used:

- *Model complexity*, which refers to the number of control points of the B-spline model;
- *Estimation accuracy*, which is defined as the MSE (mean squared errors) between the actual data points and the reconstructed model chosen by a selection criterion.

The model complexity and estimation accuracy provide insights into the appropriateness of model fitting (i.e. over-fitting or under-fitting). In the current implementation, a uniform B-spline is used for reconstructing the cross section curves whose control points are uniformly distributed in the interval between the two end points of the curve in the parameter space. In selecting the model for a cross section curve, the number of control points is iteratively incremented by one from the initial minimum number, while the corresponding BIC value is evaluated using (8.19). The minimum number of control points of a B-spline model is normally set at six here.

We first conducted experiments with only partial data of an object surface acquired by our range sensor from the first view. The object is the head of a statue as shown in Fig. 8.4. For each of the cross section curves, some data points were available for its reconstruction. Here we describe the modeling process via an example in reconstructing one cross section curve. To implement our improved BIC, the available data were first divided into two parts: a training sample set and a prediction sample set. The training sample set was used to estimate the parameters of a candidate B-spline model by (8.5), followed by the estimation of the variance σ_{kx}^2 and σ_{ky}^2 by (8.17) using the prediction sample set. The corresponding BIC value for each of the candidate B-spline models was evaluated by (8.19). The model with maximum BIC value was selected as the optimal one to approximate the data points, giving the resulting model complexity of 9. This model was then verified by using another set of data on the same cross section. The resulting curve is given in the second row in Table 8.1. The estimation accuracy, which is the mean squared errors between the actual data points and the reconstructed model, was found to be 0.0406 mm. As a comparison, the conventional BIC was also applied to the same curve. However, all the 240 data points were used in selecting the model via evaluating the BIC value by (8.19), giving the selected model complexity of 150. Again using another set of data (the same set as used in the above verification), this model was verified, with the resulting curve given in the third row in Table 8.1. The estimation accuracy in this case was found to be 1.8015 mm. This large error shows that the conventional BIC results in over-fitted approximation for the whole curve via the partial data. This illustrates the limitation of the conventional BIC criterion: its insensitivity to over-fitting. Note that in Table 8.1, the scales of the figures are set differently, in order to show the resulting errors in the reconstructed curves by different criteria which are significantly different in magnitudes. Similar phenomena were observed for other cross section curves. Here only the results for one curve are given in Table 8.1. In practical implementation, some physical constraints need to be given. For example, due to self-occlusion, the back of the object will not be visible from the first view. Some points were thus defined between the two end points of the available cross section data to limit the range of the occluded part of the object. It is useful and reasonable to confine the occluded part of the object within the range of the two end points of the available data beyond which the part would actually become visible to the current view. These defined points are highlighted in the "blue box" in the figures in the first row in Table 8.1.

The second experiment was conducted with the one where complete data of a surface were available. The procedures in reconstructing the cross section curves were the same as those in the first experiment. For each section curve, verifications of the models reconstructed by the two methods (our improved BIC and conventional BIC) were again conducted using another set of data (different from that used for reconstructing the model) on the same curve, with the results listed in the second column in Table 8.1. From the results, it is observed that even with complete data for a curve, the conventional BIC still results in an over-fitted model as seen in the large errors in the verification, while our improved BIC method can reconstruct these cross section curves satisfactorily. With more data available in this experiment, the complexities of the selected models increased using both selection criteria. Yet, the conventional BIC performed poorly with apparent over-fitting in its reconstructed models.

Table 8.1. Comparison of the results of our improved BIC with conventional BIC

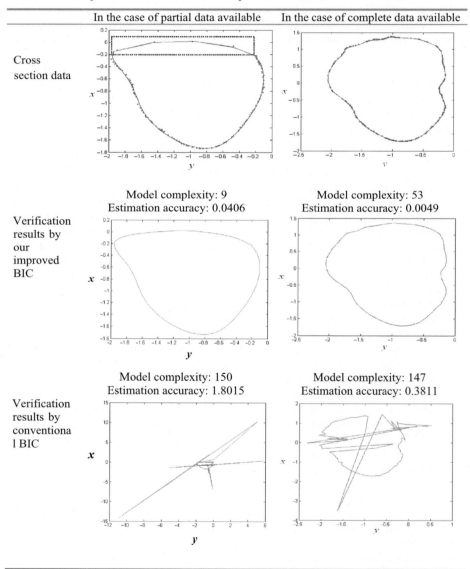

	In the case of partial data available	In the case of complete data available
Cross section data		
	Model complexity: 9 Estimation accuracy: 0.0406	Model complexity: 53 Estimation accuracy: 0.0049
Verification results by our improved BIC		
	Model complexity: 150 Estimation accuracy: 1.8015	Model complexity: 147 Estimation accuracy: 0.3811
Verification results by conventional BIC		

8.5.3 Determining the NBV

In the above section, we showed how our improved BIC criterion selects the B-spline model for the reconstruction of cross section curves. In this section, we will analyze the uncertainty of the B-spline model selected by our improved BIC for each cross section curve, and predict the information gain of the model along each curve using (8.29). Based on this analysis, we then map the information gain onto the view space. The view with maximum information gain is selected as the NBV. Then the vision sensor can take another measurement from the NBV to update the B-spline model. We will take one cross section curve as an example to illustrate the process in determining the NBV.

8.5.3.1 Determining the First NBV

First, we take the measurement from an arbitrary initial viewpoint to acquire the first part of data of the unknown object. The data points on one of the cross section curves are shown in Fig. 8.5a. The "blue box" in Fig. 8.5a contains the points to confine the range of the occluded part of the object. Since these points are few in number, their effects on the predicted information gain of the B-spline model can be ignored. Figure 8.5b is the reconstructed B-spline model using the partial data acquired from the first viewpoint. This model is a rough approximation for the whole cross section curve. Using this model, we predict the potential information along the reconstructed curve. As shown in Fig. 8.5c, the place on the curve where the data are missing (the missing part) corresponds to a high-potential information gain. This indicates that the occluded part should be given high priority in the next measurement. Note that the information gain (in Fig. 8.5c) is given in the parameter space of the B-spline curve here.

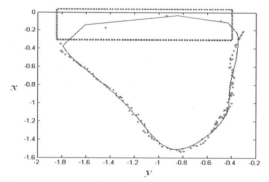

(a) Data on a cross section curve acquired from the first view

Fig. 8.5. Reconstruction of cross section curve and predicted potential information gain under the first viewpoint

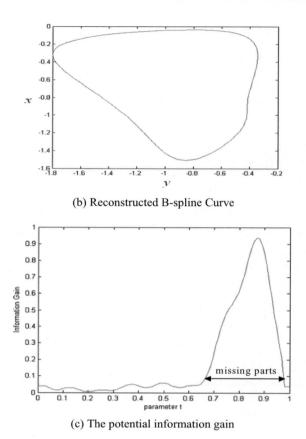

(b) Reconstructed B-spline Curve

(c) The potential information gain

Fig. 8.5. (Continued)

Following the above procedure, each cross section curve is reconstructed in a B-spline model, with the corresponding information gain obtained. Here each cross section curve is considered to be equally important, so that we can normalize the predicted information gain for each of the cross section curves covered by the current view. Figure 8.6 shows all the cross section curves reconstructed from the 3D data points taken from the first viewpoint.

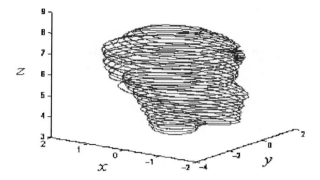

Fig. 8.6. The reconstructed cross section curves

In the above reconstruction, since only the data points from the first viewpoint are available, the obtained B-spline model cannot describe the whole object accurately. Yet, it enables us to obtain a rough shape and the information gain about the object. Based on the reconstructed partial model, we then map the predicted information gain onto the view space. As a result, we can obtain the relationship between the predicted information gain about the object and the viewpoints, which is also referred to as "View Space Visibility". As shown in Fig. 8.7, the viewpoint at $[-3.0°, 107°]$ has the maximum information gain and is thus selected as the NBV.

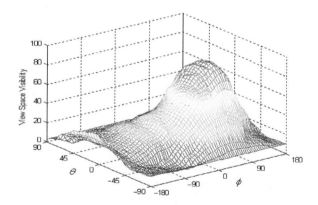

Fig. 8.7. "View Space Visibility" for the first NBV

8.5.3.2 Determining Further NBVs

After the first NBV was selected, the robot was commanded to move the vision sensor to this viewpoint to take new measurements. The newly acquired data were then sliced and registered, to yield the data acquired from the first two viewpoints as shown in Fig. 8.8a.

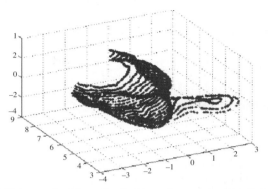

(a) Data acquired from the first two viewpoints after slicing

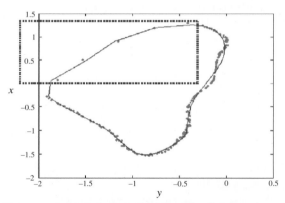

(b) Data on a cross section acquired from the first two viewpoints

Fig. 8.8. The process of determining the second NBV

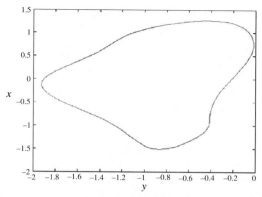

(c) Reconstructed B-spline Curve based on the first two viewpoints

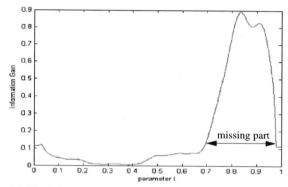

(d) The information gain based on the first two viewpoints

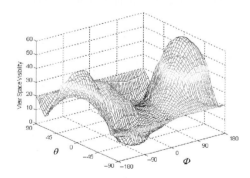

(e) "View Space Visibility" for determining the second NBV

Fig. 8.8. (Continued)

Using the available data, model selection and information gain prediction were performed following the same procedures as described above. For an example cross section shown in Fig. 8.8b, the newly reconstructed curve is given in Fig. 8.8c and the updated information gain is given in Fig. 8.8d. The predicted information gains for all the cross section curves were then mapped onto the view space, to give the updated view space visibility (shown in Fig. 8.8e) for determining the second NBV. From this view space visibility map, the second NBV was selected at [5°, 160°].

The above described procedures in determining the NBV and acquiring new data are repeated for subsequent NBVs. The procedures and results in determining the third NBVs are given in Fig. 8.9. Each time when new data are available from the new viewpoint, the corresponding cross section curves (e.g. the curve in Fig. 8.5b) are updated (as shown in Figs. 8.8c and 8.9b). The prediction of the information gain is also updated at each new viewpoint, as seen in Figs. 8.8d and 8.9c. As a result of the updated "View Space Visibility" evaluation at the second NBVs (see Fig. 8.9d), the third NBV was selected at [7°, −10°].

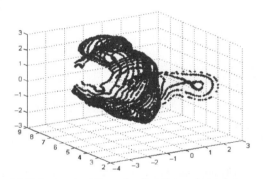

(a) Data acquired from the first three viewpoints

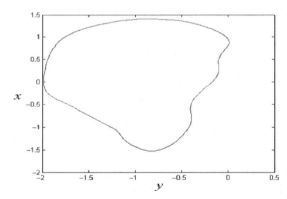

(b) Reconstructed B-spline Curve based on the first three viewpoints

Fig. 8.9. The process in determining the third NBV

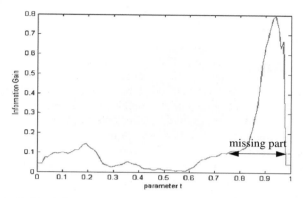

(c) The information gain based on the first three viewpoints

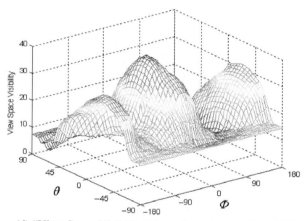

(d) "View Space Visibility" for determining the third NBV

Fig. 8.9. (Continued)

8.5.3.3 Complete Reconstruction

After the third NBV is determined, we obtained the complete data about the object as shown in Fig. 8.10a. The complete data points and final reconstruction result of a cross section curve are shown in Fig. 8.10b and c respectively.

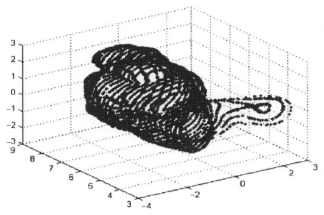

(a) Data acquired from the first four viewpoints

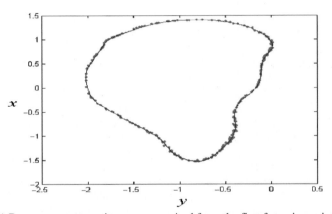

(b) Data on a cross section curve acquired from the first four viewpoints

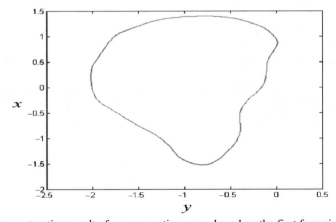

(c) Reconstruction result of a cross section curve based on the first four viewpoints

Fig. 8.10. Reconstruction of a cross section curve and information gain

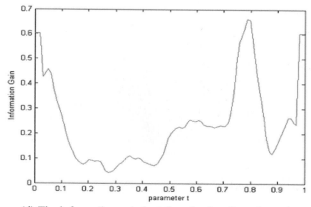

(d) The information gain based on the first four viewpoints

Fig. 8.10. (Continued)

As shown in Figs. 8.5c, 8.8d and 8.9c, the information gain has an outstanding peak on the part where the 3D data are missing. This peak will become less and less outstanding with the increase of the 3D data available from new viewpoints. When complete data on these cross section curves are obtained (as from the third NBV here), the peak in the information gain becomes non-apparent and appears more "noise" like (as seen in Fig. 8.10d), which indicates that there are no apparent missing data or occluded parts on the object surface. The disappearance of the peak (significant decrease in the peak value) in the information gain was used as the termination condition in automated planning of the NBVs.

From the experiment results, it is observed that the reconstructed model complexity tends to increase with the availability of additional data, which indicates that the model can describe the previously unknown object in more and more details as new measurements are taken. At the same time, the uncertainty about the object decreases gradually. The results for a typical cross section curve are shown in Table 8.2. The finally reconstructed model is visualized in Fig. 8.11. The final reconstruction accuracy evaluated using MSE between the actual data points and the reconstructed cross section B-spline curves was 0.0061, which is quite satisfactory.

Table 8.2. The results of view planning for the statue

Next best view	1st viewpoint	1st NBV	2nd NBV	3rd NBV
Model complexity	7	11	22	26
Entropy of B-spline model	−15.81	−16.23	−18.95	−19.79

Fig. 8.11. The finial reconstruction result of the statue

8.5.4 Another Example

Another experiment was conducted using a model of a duck. For simplicity, we only give the results (in Fig. 8.12) to show the procedures of determining the first NBV.

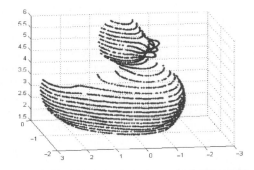

(a) Data acquired from the first viewpoint of the duck model

Fig. 8.12. Reconstruction of cross section curves and predicted information gain

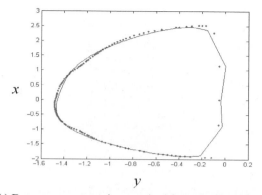

(b) Data on a cross section acquired from the first viewpoints

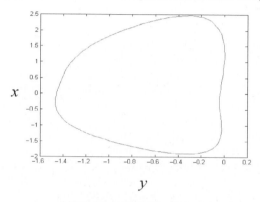

(c) Reconstructed B-spline curve

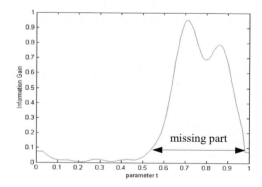

(d) The information gain

Fig. 8.12. (Continued)

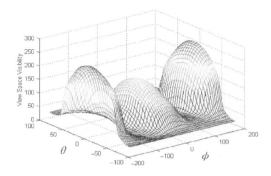

(e) "View Space Visibility" for the first NBV

Fig. 8.12. (Continued)

The viewpoint [0°, 175°] with maximum information gain was selected as the NBV. The procedures of determining other NBVs are the same as those described in the above section. In this example, three viewpoints in total were needed to reconstruct the duck model. The results in view planning for a typical cross section curve are shown in Table 8.3. The accuracy of the finally reconstructed object surface is 0.0076. The reconstructed object is shown in Fig. 8.13. It is observed that the model complexity for the finally reconstructed cross section curve (68) here is higher that that for the example curve (26) in the previous experiment. This is due to the difference in the shapes from the actual data points. The shape of the former curve (partly given in Fig. 8.12b) is simpler and smoother than the latter (Fig. 8.10b). A higher complexity in the selected model indicates the higher level of confidence in the reconstruction for a simpler shape. For a complex shape, a lower complexity in the selected model gives it stronger ability in preventing over-fitting the data, which is of particular importance for NBV planning.

Table 8.3. The results in view planning for the duck model

Next best view	1st viewpoint	1st NBV	2nd NBV
Model complexity	7	35	68
Entropy of B-spline model	−14.26	−20.23	−30.56

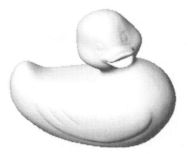

Fig. 8.13. The finial reconstruction result of the duck model

8.6 Summary

In this chapter, we presented a novel viewpoint planning method by incrementally reducing the uncertainties of the reconstructed models. With this method, the object's surface is first decomposed into a set of relatively simple cross section curves, with each to be reconstructed by a set of closed B-spline curves. Then the uncertainties of the B-spline models are analyzed with the information entropy as the measurement of the uncertainty for guiding the selection of the next best view. The information gain of the set of cross section B-spline models is predicted and mapped onto the view space. The viewpoint with maximum visibility is selected as the Next Best View. In addition, an improved BIC criterion is proposed for the model selection. With this new criterion, the acquired data points are divided into two parts: one for estimating the B-spline model parameters and the other for estimating the data noise. The re-sampling of the data enables a reliable estimate of data noise, since the generalization capability of a B-spline model should be validated using another set of data points rather than those used for the approximation. Compared with the conversional BIC criterion, the model selected with our improved BIC criterion is more sensitive to over-fitting and thus has a better generalization capability which is particularly important for NBV planning.

Chapter 9
Model Prediction and Sensor Planning

To increase the modeling efficiency, this chapter is about to present a method of viewpoint planning for incrementally building the model of an unknown object or environment. The proposed method is based on the model of the trend surface, which is the regional feature of a surface for describing the global tendency of change. Whilst previous approaches to trend analysis usually focused on generating polynomial equations for interpreting regression surfaces in three dimensions, this research proposes a new mathematical model for predicting the unknown area of the object surface. A unique surface model is established by analyzing the surface curvature. Furthermore, a criterion is given to determine the exploration direction. Algorithms are developed for determining the next view pose which needs to satisfy the sensor placement constraints such as resolution, focus, and field of view.

9.1 Surface Trend and Target Prediction

Multiple views are required to reconstruct a 3D model of a complete object or environment. Several single depth images are acquired from different views and merged together with geometric fusion techniques to produce a representation of the underlying 3D target. This is the basic idea in model-building tasks.

9.1.1 Surface Trend

In this research, the viewpoint is determined according to the surface trend of the known partial model. The *Surface Trend* describes the global shape of a surface and trend surface analysis is a global method for processing spatial data. Mathematically, a mapped surface can be separated into two components – that of the trend and the residuals from the trend. The trend is the regional feature of a surface, and the residuals are the local fluctuations of high-frequent features (Fig. 9.1a).

Trend surface analysis is widely used for fitting and interpolating regression surfaces in three dimensions as a smoothed representation of area data. It is assumed that the spatial distribution of a particular phenomenon can be represented by some form of continuous surface, usually a defined geometric function. The observed spatial pattern can be regarded as the sum of such a surface and a

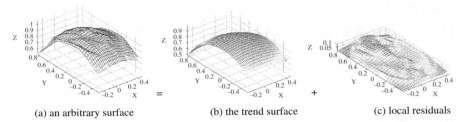

(a) an arbitrary surface   (b) the trend surface   (c) local residuals

Fig. 9.1. The trend is the regional feature of a surface

"random", or local, term. The surface is a function of two orthogonal coordinate axes which can be represented by

$$z = f(x, y) + e, \tag{9.1}$$

in which the variable z at the point (x, y) is a function of the coordinate axes, plus the error term e. This expression is the generalized form of the General Linear Model, which is the basis of most trend methods.

The function $f(x, y)$ is usually expanded or approximated by various terms to generate polynomial equations. To develop complex, smoothed equations for geophysical data by expanding the summation term of the General Linear Model (Agocs 1951, Krumbein 1959), the relationship between standard multivariate regression analyses and trend methods can be defined. This expansion was performed by incorporating power terms and cross-products of the x and y coordinates. For an n-order three-dimensional surface, the form of the power series is given by

$$f(u, v) = \sum_{i=0}^{n} \sum_{j=0}^{i} b_{ij} u^j v^{i-j}, \tag{9.2}$$

where u and v are the coordinates on an arbitrary orthogonal reference system, b_{ij} is the constant coefficient of the surface (b_{00} is the surface base).

The trend part is very helpful for predicting the unseen part of an object or environment and is thus used for determining the next viewpoint in this research. The residuals (local features) do not affect viewpoint planning much, but they should be filtered out during the image processing.

Here we split a single surface M into two parts, M_1 and M_2 (Fig. 9.1b and c),

$$M = M_1 \cup M_2. \tag{9.3}$$

If surface M changes smoothly, both the trends of M_1 and M_2 should be approximately equal to the trend of M, i.e.

$$\text{Trend}(M) \approx \text{Trend}(M_1) \approx \text{Trend}(M_2). \tag{9.4}$$

Suppose that the vision agent has already captured a part of the surface, say M_1, and M_2 is unknown. In the modeling task, by computing the surface trend of M_1, the surface shape of M_2 can be predicted. This research does not use (9.1) or (9.2) directly as the trend model for surface prediction, since it relies on interpreting regression of the known area. Instead, a new mathematical model is developed for describing the surface trend, thus emphasizing the prediction of the unknown area.

9.1.2 Determination of the Exploration Direction

Except for surface edges and object boundaries (that will be discussed later), since the curvature of a trend surface changes smoothly, the unknown part of the object surface can be predicted by analyzing the curvature tendency of the known surface. Assume that the known part is located in the center of the scene and its surrounding areas are unknown (Fig. 9.2). Since only one direction, called exploration direction, can be chosen for planning the next viewpoint, it may be determined as such an area that is most smoothed or with the lowest surface order. The surface order is determined according to $f(u, v)$ with the same fitting error. Figure 9.2 illustrates the selection criterion of the exploration direction.

With the partially acquired model, we could obtain the curvature distribution of the known surface. The unknown part of the object surface can be predicted by analyzing the curvature tendency of the known surface. As the center area of the known model does not affect the exploration direction, it is only necessary to compute the surface curvature in the area near to the boundary of the known surface. As it is difficult to directly use polynomial equations to describe the known partial surface due to its unknown complexity, the known surface should be segmented according to the curvature distribution. Then the surface regions that

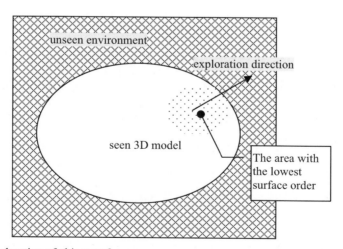

Fig. 9.2. Exploration of object surface

join smoothly at their common boundaries are merged to create the final surface region description. Here, to avoid surface fitting, we will compute the Gaussian curvature and the mean curvature from its adjacent triangles (Razdan and Bae 2003) to be later used to classify the surface type.

The Gaussian curvature K at a vertex point is computed by

$$K = \frac{\rho \Delta \theta}{A}, \text{ with } \Delta \theta = 2\pi - \sum_i \theta_i, \text{ and } A = \sum_i A_i$$

where A is the total area of the adjacent triangles T_i, for $i = 1, 2, 3, \ldots,$ and ρ is a constant (3 here). Figure 9.3 shows an example of curvature approximation at vertex P_0.

The mean curvature is defined by the divergence of the surface around the normal vector, $H = \nabla \vec{n}$. The mean curvature normal for a surface mesh is computed as (Schneider and Kobbelt 2001)

$$- H\bar{n} = \frac{1}{4A} \sum_{j \in N(i)} (\cos \alpha_j + \cos \beta_j)(P_j - P_i)$$

where $N(i)$ is the vertex P_i's adjacent polygon set, $(P_j - P_i)$ is the edge e_{ij}, α_j and β_j are two angles in the $(j+1)$th and $(j-1)$th element in $N(i)$ opposite to the edge e_{ij}, respectively. A is the sum of the areas of triangles in $N(i)$. Figure 9.4 shows the approximation of the mean curvature at a vertex.

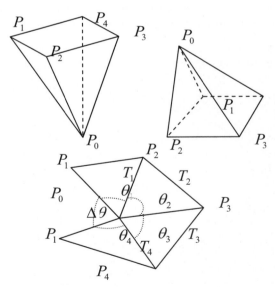

Fig. 9.3. Approximation of Gaussian curvature at vertex P_0

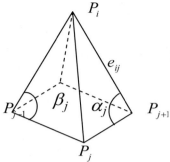

Fig. 9.4. Approximation of mean curvature at vertex P_i

Suppose M is the mesh structure of object O, and M consists of two lists V and E.

$$V = \{v_i | i = 1,\ldots, N_v\}$$

$$E = \{e_i | i = 1,\ldots, N_e\}, e_i = \{v_p, v_q\}$$

where v_i are vertices, $1 \le p, q \le N_v$ and $p \ne q$. V is a list of all vertices v_i in the mesh. E is a list of edges e_i, which connect two vertices. N_v and N_e are the number of vertices and the number of edges in M, respectively.

Besl and Jain (Massions and Fisher 1998) presented a technique which used the signs of the Gaussian curvature (K) and the mean curvature (H) to segment surfaces into patches which were labeled as belonging to one of the eight fundamental types as shown in Table 9.1.

The boundary area of the known surface can be divided into different surface types according to Table 9.1. In these areas, quadrics can be fitted to each patch. Patches containing similar fits are merged, with the exception of patches located on lines of high curvature. When the patch merging is finished, the boundary number of different surface areas is determined.

Here, the next best view is defined as the next sensor pose which will acquire the greatest amount of data of the boundaries. Thus, after the number of boundaries is obtained, i.e. $n_{max} = \max\{\text{number(Surfaceboundaries)}_i\}$, the exploration direction is determined to be along the direction outside the unknown area.

Table 9.1. Surface types and curvature signs

	$K > 0$	$K = 0$	$K < 0$
$H < 0$	Peak $T = 1$	Ridge $T = 2$	Saddle ridge $T = 3$
$H = 0$	None	Flat $T = 4$	Minimal surface $T = 5$
$H > 0$	Pit $T = 6$	Valley $T = 7$	Saddle valley $T = 8$

9.1.3 Surface Prediction

It is noted that there are different curvatures for a surface point along different directions, although the principal curvature and Gaussian curvature are the most frequently used. Without loss of generality, we may describe the mathematical formula along the horizontal direction. Using a vertical sectional plane which is parallel to the x-axis, at $y = y_v$, to cut through the 3D surface, a surface curve (Fig. 9.5) is obtained,

$$z_v = f_{yv}(x).\tag{9.5}$$

The curvature of the curve is

$$k = k_{yv}(x) = z_v'' / [1+(z_v')^2]^{3/2}.\tag{9.6}$$

Let $X_k = [x_1, x_2]$ be the domain of the known part of the surface curve. To predict the unseen surface, here we use a linear regression of x on k and a fitted curve C_v is obtained for approximating the curvature tendency on curve z_v. Hence,

$$c_v(x) = a\,x + b, \ x \in [x_1, x_3],\tag{9.7}$$

where $[x_1, x_3]$ is the whole domain including the known area and unseen area, i.e. $[x_1, x_3] = [x_1, x_2] \cup [x_2, x_3]$. The two parameters a and b are fitted by the known part of the surface curve, i.e.,

$$a = \frac{6(x_2 + x_1)\int_{x1}^{x2} k(x, y_v)\mathrm{d}x - 12\int_{x1}^{x2} xk(x, y_v)\mathrm{d}x}{(x_1 - x_2)^3},\tag{9.8}$$

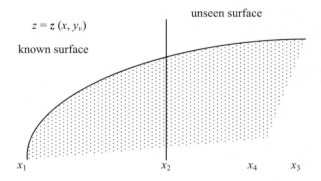

Fig. 9.5. A surface curve

and

$$b = [\int_{x1}^{x2} k(x, y_v)dx - \frac{1}{2}(x_2^2 - x_1^2)a]/(x_2 - x_1).$$ (9.9)

Given a threshold k_{max}, the curvature in the unseen area is expected to be:

$$c(x, y_v) = \begin{cases} ax + b, & c(x, y_v) < k_{max} \\ k_{max}, & c(x, y_v) \geq k_{max} \end{cases}, \qquad x \in [x_2, x_4],$$ (9.10)

where $x_2 < x_4 < x_3$ is a domain for satisfying the constraint that the object surface will be in the Field-of-View of the sensor.

Then the surface curve in the unseen part of the object will be a solution of the following equation:

$$\|z''\| / [1+(z')^2]^{3/2} - c(x, y_v) = 0.$$ (9.11)

The solution of this differential equation is:

$$z = \pm \int \frac{(ax^2 + 2bx + 2C_1)}{\sqrt{4 - (ax^2 + 2bx + 2C_1)^2}} dx + C_2, \quad x < (k_{max} - b)/a,$$ (9.12)

or

$$z = \sqrt{k_{max}^2 - (x - C_3)^2} + C_4, \quad x \geq (k_{max} - b)/a,$$ (9.13)

where C_i are the differential constants which can be determined according to boundary conditions, such as $z(x_2) = z_2$ and $z'(x_2) = z_2'$. The sign "+" or "−" can also be determined by the known part of the surface curve (convex or concave). Since the predicted curve is based on the analysis of the tendency of the known area, it is called the *trend curve*.

9.2 Determination of the Next Viewpoint

To determine the next viewpoint is to specify the sensor's placement parameters as well as to satisfy the placement constraints. The placement parameters include the sensor's position (x, y, z), the sensor orientation (α, β, γ), and some optical settings $(c_1, c_2, \ldots)$. The placement constraints usually include visibility, focus, field of view, viewing angle, resolution, overlap, occlusion, and some operational constraints such as the kinematic reachability of the sensor pose and robot-environment collision.

Let the resolution constraint be

$$r_p = 2\sqrt{(x_p - x_2)^2 + [z_p - z(x_2, y_v)]^2} \tan\left(\frac{\Omega}{2}\right)/N < r_{max},$$ (9.14)

where N is the pixel number on a scanning line of the digital image and Ω is the angle of view.

To satisfy the constraints of sensor placement on the resolution and the Field-of-View, the parameter x_4 in (9.10) can be determined by the following algorithm:

1. Obtain the numerical solution of (9.12) or (9.13);
2. Assign the z-value in an array $A_z[i]$, $i=1, 2, 3,..., n$, $A_z[1]=z(x_2, y_v)$; the corresponding x-value is $A_x[i]=x_2 + (i-1)*w_x$, where w_x is the pixel length in x-direction;
3. For $(x_p=x_3, x_p>x_2, x_p=x_p-x_{step})$,
4. If r_p satisfies the constraint (9.14), break;
5. Let $x_4 =x_p$

The mid-point of such a trend curve is:

$$\mathbf{Q}_{m,yv}= (x_v,y_v,z_{m,v}), \ z_{m,v} = f(x_v,y_v), \ x_v = \frac{x_2 + x_4}{2}. \tag{9.15}$$

By moving the $V-V$ plane to different positions, in the domain of $-Y_{fov} < y_v < +Y_{fov}$, we get a series of surface curves. Connecting the mid-point of each such curve forms a new curve:

$$\mathbf{L}_{i+1} = L(\mathbf{Q}_{m,yv} , y_v), \ -Y_{fov} < y_v < +Y_{fov}. \tag{9.16}$$

Computing its centroid, the position of the reference point (i.e. the new scene center) is obtained:

$$\mathbf{O}_{i+1}: \ x^o_{i+1} = \frac{\int_{L_{i+1}} xLdl}{\int_{L_{i+1}} Ldl}, \quad y^o_{i+1} = \frac{\int_{L_{i+1}} yLdl}{\int_{L_{i+1}} Ldl}, \quad \text{and } z^o_{i+1} = \frac{\int_{L_{i+1}} zLdl}{\int_{L_{i+1}} Ldl}, \tag{9.17}$$

where $(i+1)$ denotes the next view pose.

Now the position of the eye point and the viewing direction can be determined. To achieve the maximum viewing angle (i.e. the angle between the viewing direction and the surface tangent) for minimizing the reconstruction uncertainty, the viewing direction is the inverse of the average normal on the predicted surface, that is,

$$\mathbf{I}_{i+1} = -\frac{\iint N(x,y)dxdy}{\iint dxdy} = -\frac{\iint(\mu\mathbf{i},v\mathbf{j},\kappa\mathbf{k})(x,y)dxdy}{\iint dxdy} \tag{9.18}$$

$$= (-\mu_{i+1}\mathbf{i}, -v_{i+1}\mathbf{j}, -\kappa_{i+1}\mathbf{k}),$$

where $\mu(x,y) = \dfrac{\partial f}{\partial x}$, $v(x,y) = \dfrac{\partial f}{\partial y}$, $\kappa(x,y) = -1$, and $N(x, y)$ is the surface normal on point (x, y, z).

The sensor's position $P_{i+1}(x, y, z)$ for the next viewpoint is planned as a solution of the following set of equations:

$$
\begin{cases}
\dfrac{x^o_{i+1} - x^P_{i+1}}{\| O_{i+1} - P_{i+1} \|_2} = \dfrac{-\mu_{i+1}}{\| I_{i+1} \|_2} \\[3mm]
\dfrac{y^o_{i+1} - y^P_{i+1}}{\| O_{i+1} - P_{i+1} \|_2} = \dfrac{-v_{i+1}}{\| I_{i+1} \|_2} \\[3mm]
\| O_{i+1} - P_{i+1} \|_2 - \dfrac{r_{max} N}{2 \tan \dfrac{\Omega}{2}} = c_{cmp}, \quad c_{cmp} \geq 0
\end{cases}
\tag{9.19}
$$

where c_{cmp} is a positive constant for compensation of the depth value range, Ω is the sensor's Angle-of-View, $\| I_{i+1} \|_2 = \sqrt{\mu_{i+1}^2 + v_{i+1}^2 + \kappa_{i+1}^2}$, and $\| O_{i+1} - P_{i+1} \|_2$ is the distance between point O_{i+1} and P_{i+1}.

Using this method, the orientation of the sensor can be determined in such a way that it views perpendicularly downon the predicted object surface. The distance to the object is determined so that it enables the acquisition of the maximum volume of the unknown surface while satisfying some constraints such as resolution. Finally, the placement parameters of the vision sensor are described as a vector:

$$
P_{i+1} = (x^P_{i+1}, y^P_{i+1}, z^P_{i+1}, \mu_{i+1}, v_{i+1}, \kappa_{i+1}, c^1_{i+1}, c^2_{i+1}, ..., c^n_{i+1}),
\tag{9.20}
$$

where $\{c_{i+1}^j\}$ are the optical settings of the sensor for the next viewpoint, such as focus and diameter of aperture, which may be different depending on the types of vision sensor. Also this placement vector is based on the local coordinate system. It needs to be converted to the world coordinate system by multiplying a coefficient matrix.

It should be noted that the surface predicted by the trend is only the possible surface shape. This research conducted an error analysis, under the assumption that the actual surface is composed of typical 1st or 2nd order curves. Results show that there is no error for predicting circular curves or linear curves and the errors are very slight (relative error is at the order of 10^{-7}) for predicting other 2nd order curves, such as parabola, elliptic, hyperbola. Therefore we can accurately place the vision sensor for the next view pose to observe objects composed of such surfaces, when using the expected curve predicted by the uniform surface trend model (10^{-4}). However, for each different curve the constants a, b, and C_1 are different and they can be dynamically determined according to the known part of the curves.

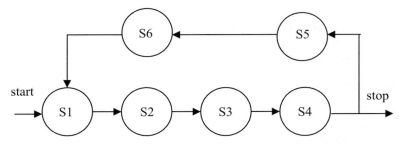

Fig. 9.6. The repetitive modeling process

Finally the repetitive modeling process is illustrated in Fig. 9.6, where the six symbols Si ($i = 1, 2, ..., 6$) represent:
- S1: Acquisition of a view
- S2: Reconstruction of the 3D local model
- S3: Registration and fusion with the global model
- S4: Model analysis and check of complete conditions
- S5: Computing trend surface and determining next view
- S6: Moving the robot to the new viewpoint

9.3 Simulation

9.3.1 Practical Considerations

9.3.1.1 Noise Filtering

Since the curvature on the object surface is very sensitive to noise and the local features of a 3D image do not affect the surface trend very much, a low-pass filter is applied to the image so that we can obtain a smoothed surface. In this research, a 5-point averaging filter is used to remove the local features of the image:

$$H = [1\ 1\ 1\ 1\ 1]/5 \text{ or ones}(5, 5)/25. \tag{9.21}$$

Applying this filter several times to the surface curve or image, the smoothed range image is used to compute the trend surface.

9.3.1.2 Determining the Curvature

The curvature of each point on the known part of the object surface should be calculated so that the surface trend can be determined for predicting the unknown part. Equation (9.6) is not suitable for computing the curvature on a digital image because the errors in computing z' and z'' will be significant. In this research, we determine the curvature of a point $P(x(i), z(i))$ by using three adjacent points (triplet), i.e $P(x(i-1), z(i-1))$, $P(x(i), z(i))$, and $P(x(i+1), z(i+1))$. Each triplet defines a circle and the curvature $k(i)$ is the inverse of its radius. Referring to (9.5), this is the solution of

$$[k(i) * (x(n) - x_0)]^2 + [k(i) * (z(n) - z_0)]^2 = 1,$$
$$n = \{i-1, i, i+1\}$$
(9.22)

9.3.1.3 Determining Constants a and b

Equations (9.8) and (9.9) are described as continuous functions. When applied to digital image processing, they may be written as,

$$a = \frac{m\sum_{i=1}^{m} x_i k_i - \sum_{i=1}^{m} x_i \sum_{i=1}^{m} k_i}{m\sum_{i=1}^{m} x_i^2 - \left(\sum_{i=1}^{m} x_i\right)^2}$$
(9.8')

$$b = \left(\sum_{i=1}^{m} k_i - a\sum_{i=1}^{m} x_i\right) / m$$
(9.9')

where m is the number of total points of the known part of object surface.

9.3.2 Numerical Simulation

9.3.2.1 Next Viewpoint to Look at a Regular Object

Figure 9.7 illustrates an object to be reconstructed by computer simulation. It is digitally generated by rotating a generatrix about the z-axis:

$$y = 2 + \cos(t), \qquad t \in [-1.2\pi, \ 1.2\pi].$$

Assume that the sensor's field-of-view has an angle of $\Omega = 40°$ and the pixel number on a scan line is $N = 1024$. Given that the first view is located at $P_1 = [3.4964 \ \ 1.6537 \ \ 0.5417]$ and the resolution constraint is $r_{max} = 0.85$ mm. Then the surface trend is computed and the next viewpoint is determined. The result is:
viewpoint_first $= [x \ \ y \ \ z \ \ ai \ \ bj \ \ ck]$
    $= [\ 3.4964 \ \ \ 1.6537 \ \ \ 0.5417 \ \ \ 0.9040 \ \ \ 0.4276 \ \ \ \ \ 0\ ],$

viewpoint_next
    $= [0.7691 \ \ \ 3.7241 \ \ \ 0.5417 \ \ \ 0.1951 \ \ \ 0.9808 \ \ \ \ \ 0\ ].$

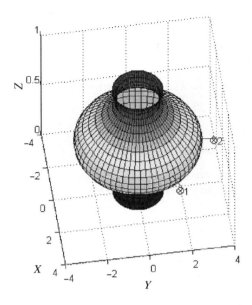

Fig. 9.7. Planning the next viewpoint for modeling a regular object

9.3.2.2 Planning Next Viewpoint for Freeform Objects

To show that the method can also work well for the modeling of freeform objects or environments, an example is given to compute the next viewpoint for modeling a torso. The range data were acquired by a gray-encoded stripe-projection method. In this example, the sensor's field-of-view has an angle of $\Omega = 40°$, the pixel number on a scan line is $N = 1024$, and the resolution constraint is r_{max}=0.85 mm. Figure 9.8 is a 3D image to be considered as the first view during the modeling process. To determine a next view for acquiring unseen information on the object, we used the trend surface method and developed a program to compute the expected surface curves. Thus the surface trend is computed and the next viewpoint is determined. The computation result is:

viewpoint_first is at

   [429.0327 166.1754 500.0000   0.5211   0.8535   0],

viewpoint_next is planned at

   [542.5648 −36.9134 500.0000   0.9963   0.0865   0].

Figure 9.9 illustrates a profile of the known partial model at $y = 268.0$. The red curve is the expected curve on the unknown part predicted by the surface trend model. Figure 9.10 illustrates the three-dimensional model at the first view and the next view.

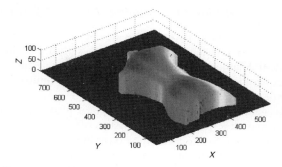

Fig. 9.8. The first view of a torso

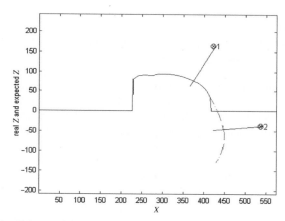

Fig. 9.9. A profile of the partial model

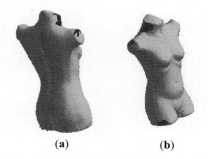

(a)   **(b)**

Fig. 9.10. The first view **(a)** and the next view **(b)** of the model

9.3.2.3 Viewpoint Planning for Modeling Unknown Environments

A computer simulation program was also developed for evaluating the vision performance. The user interface of the software is illustrated in Fig. 9.11. This simulation system enables a person to describe user-defined vision sensors, user-defined environments, 3D image acquisition, multi-view fusion, and viewpoint control.

Two models should be given before viewpoint planning. One is the vision sensor. The user may describe the characteristics of the 3D sensor in a script data file, including the initial position and the constraints of the viewing angle, focal length, zoom range, resolution, etc. A point light source may also be defined. In the current stage, this can only be fixed at a 3D point and its illumination intensity cannot be controlled. The other initial model is the environment. This may be imported from a 3D CAD file or described by scripts too. For scripts, the environment will be composed of many simple 3D objects, such as boxes, balls, cylinders, free-form 3D surfaces, etc. Bitmap texture may be assigned to "paint" on these object surfaces.

The task description gives the required conditions, such as the overlap width for view merging and the required reconstruction resolution (e.g. 0.1 mm/pixel in x, y, z). The current view generation is based on the current sensor position and orientation. It simulates the currently seen scene and generates a 3D image according to the vision sensor's description. The left window in Fig. 9.11 illustrates a current view. The function of 3D reconstruction of the seen portion fuses all seen areas of the environment into a 3D partial model to illustrate what has been reconstructed (the mid-upper window in Fig. 9.11). The "manual viewpoint control" is used for debugging purposes. It enables the user to set the parameters of the sensor pose and shows the corresponding view.

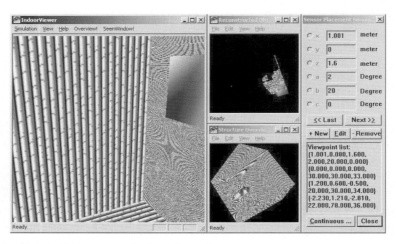

Fig. 9.11. The simulation system (IndoorViewer 1.0)

The other important function is the "autonomous exploration". This is realized based on the trend surface method proposed in this research. In all stages, there is a stack to record what has been seen. By analyzing the reconstructed and fused global model, the planner automatically decides where to look next for exploring unseen areas. Finally after several steps, the whole environment (the mid-lower window) is reconstructed as a 3D model. The generated viewpoint list is illustrated in the right window in Fig. 9.11.

9.4 Practical Implementation

Several experiments were carried out in our laboratory on the construction of object models. The range data are obtained by a structured light system set up in this research, which mainly consists of a projector and a camera. The projector is a palm-sized digital projector, PLUS U3-1080. It is connected to a computer and is controlled to generate some gray-encoded stripe-light patterns for 3D reconstruction. The CCD camera (PULNIX TMC-9700) has a 1-inch sensor and a 25 mm lens. A fixture is designed for mounting the structured light system on the end-effector of a 6DOF robot (STAUBLI RX-90B) with ±0.02 mm repeatability. This enables the 3D sensor to be freely moved to an arbitrary position in the workspace.

Figure 9.12 illustrates two objects for the model construction in the experiment. In both cases, we set the resolution constraint to be r_{max}=0.85 mm. The first one is demonstrated with the procedure in more detail here. It was incrementally built by four views. The first view is assumed to be taken from the top view. To determine a next view for acquiring some unseen information on the object, we used the trend surface method and developed a program to compute the expected surface curves. Then the trend is computed and the next viewpoint is determined.

The experimental results in the incremental construction of the first object are shown below to illustrate the computation at each step. A new surface was acquired

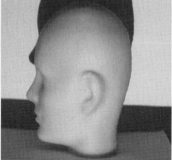

Fig. 9.12. The objects to be reconstructed

at each view and it was integrated with the existing ones to form a partial model. The exploration direction and sensor placement were determined by the proposed method. The placement parameters of each view are set with a viewpoint vector in the format of $[x \ y \ z \ a\mathbf{i} \ b\mathbf{j} \ c\mathbf{k}]$, representing the six parameters of the 3D position and orientation. A registration vector also contains six parameters for the surface integration so that the range images are transformed and registered to a common coordinate system. The first three of the six parameters define the surface's reference point and the other three define the surface orientation by 3-axis rotation. In this way, the model is improved by eliminating the overlapped/redundant points and stitching non-data areas.

The corresponding placement parameters of the first view (Fig. 9.13) are
$$\mathbf{Viewpoint}_1 = (0, 0, 457.9333, 0, 0, -1)$$
$$\mathbf{Reg}_1 = \mathbf{O}_{global} = (0, 0, 0, 0, 0, 0)$$

where the viewpoint vector has a format of $[x \ y \ z \ a\mathbf{i} \ b\mathbf{j} \ c\mathbf{k}]$, representing the six parameters of the 3D position and orientation. $\mathbf{Reg}_i$ contains some registration parameters for the surface integration so that the range images are transformed and registered to a common coordinate system. The first three of the six parameters define the surface's reference point and the other three define the surface orientation by 3-axis rotation. In this way, the model is improved by eliminating the overlapped/redundant points and stitching non-data areas.

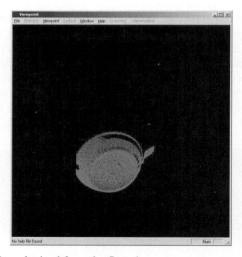

Fig. 9.13. The 3D surface obtained from the first view

Based on the partial model, surface boundaries are detected to locate candidate points (Fig. 9.14), and the rating evaluation is performed to choose a best exploration direction. Along that direction, a trend curve (the dashed red curve in Fig. 9.15) is computed from the known 3D surface. The space under the curve and ground plane is marked as unreachable.

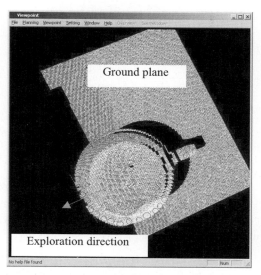

Fig. 9.14. The points were detected as the candidates of exploration directions. The decision was based on their rating values

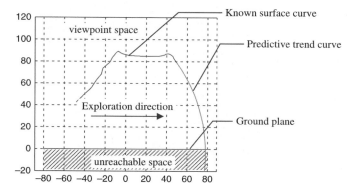

Fig. 9.15. An predictive trend curve

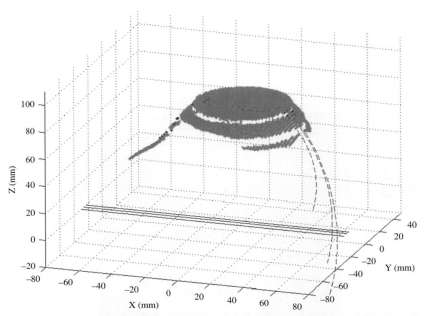

Fig. 9.16. Several trend curves are obtained to form a trend surface so that the decision of a next viewpoint will be more reliable

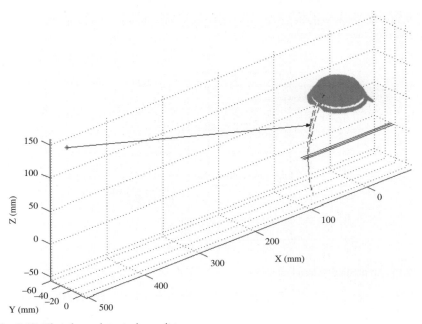

Fig. 9.17. The planned next viewpoint

The sensor position and looking direction are planned according to the predicted curves. The corresponding placement parameters are:

Set_scene_center = [63.9950, −23.0346, 40.4942].

Viewpoint2 = (502.9636, −26.2364, 158.2319, −0.965838, 0.00704474, −0.259052).

Reg2 = (−72.06, 0.00, 40.10, 32.01, −89.89, −30.80).

Similar to the situation in the first step, candidate points are selected to find an exploration direction. Along this direction, several trend curves are computed to predict the unknown part of the surface.

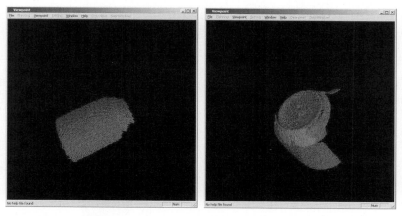

Fig. 9.18. The 3D surface was obtained from the second view (*left*) and integrated with the first view to form a partial model (*right*) of the object

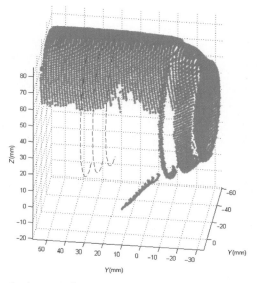

Fig. 9.19. Trend curves in the second step

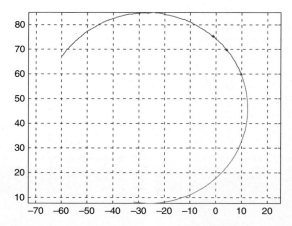

Fig. 9.20. One of the trend curves as in Fig. 9.19. Since the original curve is circular, the prediction is very accurate

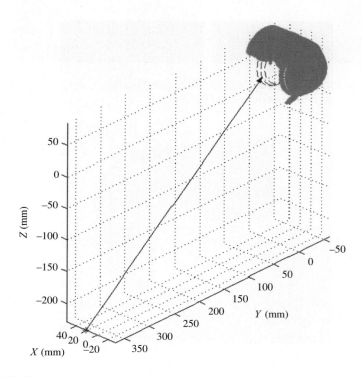

Fig. 9.21. The next viewpoint

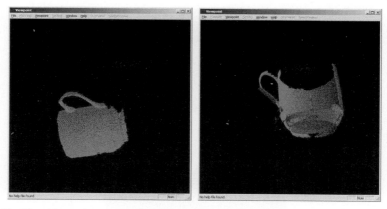

Fig. 9.22. The 3D surface was obtained from the third view and integrated with all known views

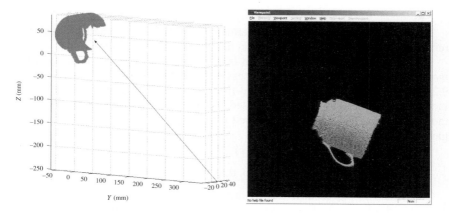

Fig. 9.23. The next viewpoint and 3D surface obtained

The sensor pose is also determined according to the predicted curves. The corresponding placement parameters are:

Set_scene_center = [−8.4859   60.0199   33.7807]

Viewpoint3 = (21.7862, 366.7046, −228.8904, −0.0747589, −0.757378, 0.648683)

Reg3 = (30.21, −116.07, 40.02, 89.83, 31.10, −89.92)

Figures 9.16 to 9.21 illustrate two steps of sensing decisions with the trend surfaces. With the model in Fig. 9.22, it is necessary to take a further viewpoint decision and surface acquisition. These were done similarly to the previous steps. The sensor parameters for the fourth view are determined as (Fig. 9.23):

Set_scene_center = [16.3405, −10.1124, 35.0603]

Viewpoint4 = (15.8901, 347.64, −251.11, 0.0009831, −0.780902, 0.624653)

Reg4 = (−116.12, −120.01, 43.03, −90.00, 22.60, 89.31)

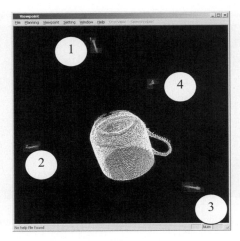

Fig. 9.24. The complete model and planned viewpoints

Finally the complete model was obtained by integrating all the four views. Their relative distribution to the object model in 3D space is given in Fig. 9.24.

The second example in this section is to show the 3D reconstruction of a head model. The entire reconstruction process was finished with 5 acquisitions. Figure 9.25 illustrates the 3D image captured from the first view of the object. The first view was always assumed to be sensed at the top-view. Assume that the object model has the original pose and the first view is registered at the same pose as the global model, i.e.

$$\mathbf{O}_{\text{global}} = (x, y, z, a, b, c) = (0, 0, 0, 0, 0, 0).$$

$$\mathbf{Reg}_1 = \mathbf{O}_{\text{global}} = (0, 0, 0, 0, 0, 0).$$

The sensor is placed at a viewpoint relative to the target and its pose is also represented by a six-dimensional vector (three for position and three for orientation). In this experiment, it is

$$\mathbf{Viewpoint}_1 = (0, 0, 457.9333, 180, 0, 0).$$

This indicates that the sensor was placed at the topside of the object with a distance z = 457.9333 mm, looking downon the target. The self-rotation angle is assumed to be $c = 0$.

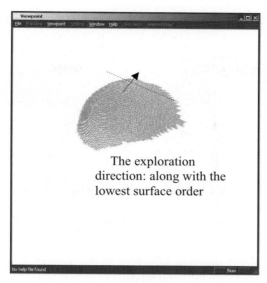

Fig. 9.25. The first view (top view) of the object

With the partial model, the pose of a following view needs to be decided according to the known information. Here two steps are used. The first step is to determine the exploration direction and the second to determine the sensor pose in the space.

With the description in Sect. 9.1.2, the exploration direction can be determined simply to be with the area where the surface is most smoothed or the surface order is lowest. The reason is that the trend surface can predict the unknown area accurately where the surface has a low order. The surface order is defined according to (9.2) with the same fitting error. To avoid computation of surface fitting, we may just compute the integral value of the curvatures in a small area, i.e.

$$n_{\text{order}}(u, v) = \iint_{x,y \in S(u,v)} k_{\min}(x, y)\mathrm{d}x\mathrm{d}y, \tag{9.23}$$

where $S(u, v)$ is the neighborhood area of point (u, v) and $k_{\min}(x, y)=f'' / [1+(f')^2]^{3/2}$ is the minimum curvature at point (x, y).

It is only necessary to compute the surface orders in the areas near to the boundary of the known surface. The surface order in the center area of the known model does not affect the exploration direction. After the minimum surface order is obtained, i.e. $n_{\min} = \min\{ n_{\text{order}}(u, v)\}$, the exploration direction is decided just to be along outside the direction to the unknown area.

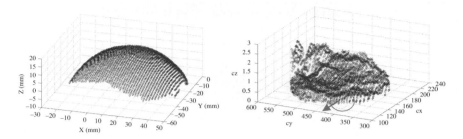

The area with the lowest surface order

Fig. 9.26. The 3D data of the first view and the distribution of local surface orders

In this experiment, the 3D data of the first view was visualized in Fig. 9.26 (left) and the surface smoothness was computed as shown in Fig. 9.26 (right). The exploration for the second view was determined as shown in Fig. 9.25. Then the next viewpoint was determined according to the method as in Sect. 9.2, i.e. the surface was assumed with the trend model

$$z = \pm \int \frac{(ax^2 + 2bx + 2C_1)}{\sqrt{4 - (ax^2 + 2bx + 2C_1)^2}} dx + C_2$$

and the sensor pose was determined according to (9.18) and (9.19). In this experiment, it is

Viewpoint$_2$ = (−128.41, 401.67, −364.82, 57.8349, 10.5665, −32.7017).

Here we did not consider the environment constraints. Then a second view was captured. The 3D surface is illustrated in Fig. 9.27. The local model is registered with the parameters being

Reg$_2$ = (−44.43, 20.59, −125.17, −57.8349, −10.5665, 32.7017).

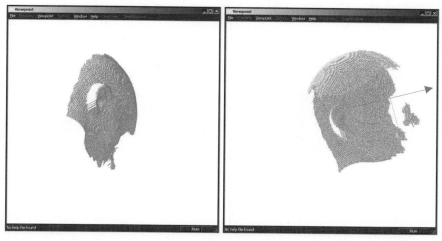

Fig. 9.27. The 2nd view of the object and merged from all knowns (the 1st and 2nd views)

To merge it with the global model $\mathbf{O}_{global}$, the surface is transformed by the following equation

$$\mathbf{O}_{2w} = \mathbf{Tr}\ \mathbf{O}_2,$$

where the 3D transformation matrix $\mathbf{Tr}$ is computed from $\mathbf{Reg}_2$

$$\mathbf{Tr} = \begin{bmatrix} 0.827225 & -0.156990 & -0.539493 & 27.5397 \\ 0.531104 & 0.531845 & 0.659698 & -95.2064 \\ 0.183376 & -0.832163 & 0.523333 & -90.7871 \\ 0 & 0 & 0 & 1 \end{bmatrix}.$$

Figure 9.27(right) illustrates the global model merged from the first view and the second view. Then a next view needs to be decided on again.

Figure 9.28 illustrates the third surface obtained and merged with previous known images. The viewpoint and model registration parameters are

$\mathbf{Viewpoint}_3 = (-414.33, 60.41, -172.36, -63.6737, 78.4519, 21.0845).$

$\mathbf{Reg}_3 = (34.33, 142.57, -131.70, 63.6737, -78.4519, -21.0845).$

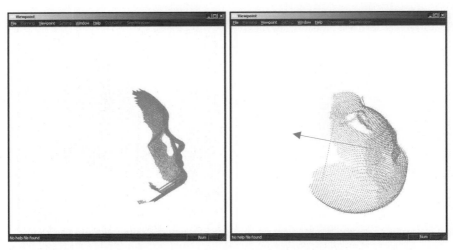

Fig. 9.28. The new surface obtained and merged with the 3 known views

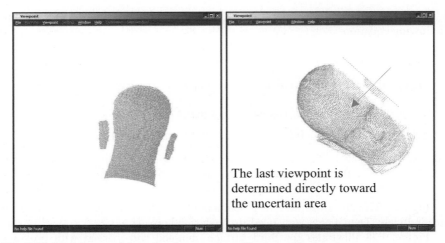

The last viewpoint is determined directly toward the uncertain area

Fig. 9.29. Model merged from four known views (the 1st – 4th views)

Figure 9.29 illustrates the fourth view of the object. The viewpoint and model registration parameters are

Viewpoint$_4$ = (531.43, 111.47, –130.98, 86.2575, –81.9966, –89.2484).

Reg$_4$ = (77.96, 47.85, –126.82, –86.2575, 81.9966, 89.2484).

The known images were merged as illustrated in the right side of Fig. 9.29. By checking the occlusion status, it was found there was only a small uncertain area. Then the last viewpoint was determined directly toward on that area.

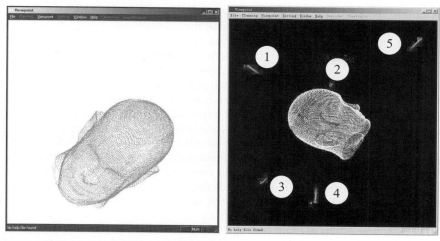

Fig. 9.30. The entire model merged from all known views and the traveled viewpoints

Viewpoint$_5$ = (112.40, −286.57, −247.88, −75.5537, −9.3295, 52.5984).

The last surface was obtained and the registration parameters are

Reg$_5$ = (38.16, 151.02, −135.15, 75.5537, 9.3295, −52.5984).

Finally the entire object model was reconstructed from all known surfaces and illustrated in Fig. 9.30.

Figure 9.30 shows the entire model obtained and all the five viewpoints traveled in the working space. In our experiments, the 3D surface acquisition was achieved by a structured light vision sensor. The computation time for planning a next viewpoint was about three to five seconds. It should be noted that the planning results are dependent on the first view. With different initial viewpoints, the results are also different and there may be one or two more views planned for each object.

9.5 Discussion and Conclusion

9.5.1 Discussion

This chapter addresses the task of reconstructing a complete 3D object, which requires the sensor to be moved (or relatively the object to be replaced) at several different poses. It also requires the exploration strategies. We adopt the trend-surface approach to this problem. This is a target-driven approach that attempts to analyze the physical structure of the object and determine where to look at the next pose for efficient reconstruction of the entire object.

For surface edges and object boundaries, as the curvature around these areas changes abruptly, there will be significant errors in computing the surface trend. For example, Fig. 9.31 illustrates that there is a surface edge located at x_2 in the seen area. In such cases, the domain of the known part for determining constants a and b should be restricted to a smaller area that does not contain any edges, e.g. using domain $[x2, x3]$ or $[x2, x1]$ instead of the whole domain $[x1, x3]$. The exploration direction may also need to be changed by analyzing the seen object surface.

On the other hand, when modeling a small-sized object (compared with the sensor's field-of-view), the trend surface method may not work very well since the whole object will be contained in a single view and the surface trend will have a high order. That makes it unreliable in predicting the unknown area. In such cases, a sensor-based solution (Zha 1998, Pito 1999) instead of target-driven method may be used for the modeling task.

The basic idea of this research is to find any possible cues on the object shape for predicting unknown areas. The surface trend is an option that is suitable for many objects. Here, the trend may not be computed by all areas of the known part. We only choose a suitable surface part usually without object boundaries or edges inside, so that the trend is reasonable and predictable. For example, for a polyhedral object, the surface trend will be a single surface plane. Only when the image boundary is on an edge of a polygonal plane, the next view can not be reasonably planned. This will certainly cause a not good viewpoint, but rarely happens in practice. Many cases are avoided by our selection method of the exploration direction. Of course, we do not expect that the trend method alone is enough to complete the whole modeling task. We have to integrate many other methods together so that the whole modeling process is executed in an "adaptive" way. That is also why we currently have difficulties to provide some complicated examples that are performed fully automatically.

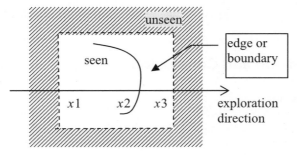

Fig. 9.31. Avoid edges or boundaries in computing the trend

9.5.2 Conclusion

In this chapter, for the nonmodel-based object modeling tasks, a planning strategy was proposed to determine where to look next for a vision agent. A trend surface model was used to predict the unseen part of unknown objects or environments. In this way, the next viewpoint can be determined for on-line acquisition of the range data until the whole structure of the object or environment is reconstructed. The trend surfaces and trend curves were computed from the curvature tendency. While determining the next viewpoint, multiple sensor placement constraints were considered in this research, such as resolution, field-of-view, and viewing angle. The error analysis shows that the trend model can accurately predict the unknown surface of the object if the surface is composed of 1st-order and 2nd-order curves and surfaces. Numerical simulation and practical tests were carried out to demonstrate the technique presented in this research.

The trend surface method works well in determining the next viewpoint for most smooth objects. However, this method alone is not adequate for efficient reconstruction. In real applications, other important techniques also have to be involved, e.g. the decision on the exploration direction should also consider surface uncertainties, occlusions, etc.

A future work in this field may deal with reliable detection of the boundary positions on a complex partial model, so that they will be evaluated for selecting a best candidate of exploration directions. Other uncertain conditions will also be considered to make the reconstruction process more reliable so that an autonomous robot system can work without any human interference.

Chapter 10
Integrating Planning with Active Illumination

The performance of the vision perception of a robot, and thus the quality of knowledge learnt, can be significantly affected by the properties of illumination such as intensity and color. This chapter presents strategies of adaptive illumination control for robot vision to achieve the best scene interpretation. It investigates how to obtain the most comfortable illumination conditions for a vision sensor. In a "comfort" condition the image reflects the natural properties of the concerned object. "Discomfort" may occur if some scene information is lost. Strategies are proposed to optimize the pose and optical parameters of the luminaire and the sensor, with emphasis on controlling the intensity and avoiding glare.

10.1 Introduction

Traditional methods for machine vision to better interpret scenes are usually focused on post- image processing (e.g. smoothing, filtering, masking, zooming, contrast stretching, pseudocoloring, etc.). However, post- image processing does NOT increase the inherent information content, but an originally better image contains more information of object surfaces. This facilitates further vision analysis and saves time-consuming enhancement processing, which is very important in a machine vision system, especially for real-time applications.

The provision of adequate light for various human activities has been a matter of some importance since the emergence of civilization. The basic question that has faced the lighting designer is "how much light do we need to see". The IES has dealt with this central problem since its beginning in 1906. In 1937, and again in 1959, a different approach was taken, based on improved understanding of the basic processes involved in vision, seeing, and performing visual tasks. The resulting work formed the basis of the illuminance recommendations of the CIE and for many years was the foundation of the IES method of prescribing illumination.

Much in the same problem, in the computer vision field, while some of the vision systems mentioned by other researchers previously have explicitly dealt with the planning of lighting parameters, current work in illumination planning is quite restricted. It should be recognized, however, that the problem of planning general lighting for machine vision is extremely difficult. Most of the work has used point sources of light that are incident on convex Lambertian surfaces. These models, while useful, are not analogous to actual lighting conditions seen in current applications. Higher-order lighting/reflecting models that include such phenomena as multiple sources (both point and extended), specularity, and interreflections from concave surfaces need to be found to properly plan lighting parameters.

The light source for a natural scene is its illumination. As with any light, illumination has the properties of intensity and color, which significantly affect the performance of the robot vision perception as well as human perception. This chapter mainly considers the most widely used "robot eye" – the CCD camera.

"As in real estate where the key to successful investments is location, location, location, in machine vision the key to value (equal to success) is lighting! lighting! lighting!" said Nello Zuech, the President of Vision Systems International.

The principal reason for success in machine vision is the elimination of appearance variables and the consistent appearance that appropriate, application-specific lighting yields. Unlike the early days of machine vision when many of the entrepreneurial researchers in pioneering machine vision companies suggested, "We just need an image, our image processing and analysis algorithms will work for your application," today people acknowledge the importance of lighting and scene consistency.

The light source for a natural scene is its illumination. For many machine-vision applications, lighting now is the most challenging part of system design, and becomes a major factor when it comes to implementing color inspection. The uniformity and the stability of the incoming lighting are usually the common causes of an unsatisfactory and unreliable performance of machine-vision systems. As with any light, illumination has the properties of intensity and color, which significantly affect the performance of robot vision perception as well as human perception.

The selection of light sources and vision sensors constitutes the first problem in vision design. There are many different kinds of sources, including incandescent filament lamps of many kinds, short arc lamps, gaseous and solid-state lasers, fluorescent lamps, high-intensity gaseous discharge lamps, electroluminescent lamps, light emitting diodes, carbon arc lamps, etc. Most CCDs have good red (long wavelength) response, but blue response can be a problem because of absorption in the polysilicon layer that covers the sensitive area. Using back-illuminated sensors may help to avoid this problem. Furthermore, a lot of camera series are ready for industrial use. It is important to select proper parameters, such as focal length, imager size, resolution, angle of view, etc.

Then optical settings and geometrical placements of the light source and the vision sensor become another problem. To solve this, we must firstly analyze what "a perfect image for machine vision" is. A good image means that it contains maximum information about the scene so that the robot can easily understand it. The evaluation criteria of illumination conditions should be given and then the degree of "comfort" to the machine eye may be analyzed.

Effort by Eltoft and de Figueiredo (1995) is one of the earliest important attempts in illumination control, and there is other literature with some relations to this problem (Gudrun 1990, Sato and Sato 1999, Ding et al. 2000). Recently, researchers have become more aware of the subject (Muehlemann 2000, Hartmann et al. 2003, Qu 2003, Martinez-De-Dios and Ollero 2004). Their work discusses many factors of illumination conditions that affect the quality of the image.

Although Chap. 3 has listed some typical works on illumination planning for active vision, this chapter further discusses many factors of illumination conditions that affect the quality of images seen by a robot. However, in our preliminary work, intensity control and glare avoidance are emphasized and fundamental sensing settings and spatial placements are proposed for active illumination setup in practical vision systems.

10.2 From Human Vision to Machine Vision

Currently CCD cameras are still the most commonly used machine eyes because of their many advantages, although CMOS cameras are also widely used nowadays. Apart from the apparent structure and working mechanism, a machine eye works very similar to a human eye with some comparable characteristics like resolution, bandwidth, luminosity, the ability to distinguish, adaptivity, and color vision.

For resolution, the ability of human vision to perceive fine detail is called acuity that is expressed as the angle subtended by the smallest object he can discern. For gray-scale objects, this is typically about 1' (minute of arc). A typical camera has an acuity of about 4'. To the human eye's bandwidth, electromagnetic radiation in the wavelength range from 400 to 700 nm is what we know as visible light and has peak responsity at 555 nm. A typical CCD element (with photodiodes or photogates) is sensitive within a wavelength between 300 and 1100 nm and has peak responsity at 800 nm. But it is practically cut off between 400 and 700 nm using filters and is normalized to meet the human sensitivity curve.

A human observer perceives the intensity (energy level) of light as the sensation called brightness. However, the perceived brightness varies depending on the color of the light. This is quantified by a luminosity curve. Usually a video camera is designed to have a spectral response that matches the similar luminosity curve. Humans can detect dozens of levels of intensity within a scene, which is referred to as gray-scale response, and thousands of colors. Present common cameras can detect $2^8=256$ different gray levels and 2^{24} true colors.

Concerning adaptivity, it is well known that the human eye adapts to average scene brightness over an extremely wide range, as much as $10^{10} - 1$. Video cameras are designed to deal with a similar brightness range and provide gray-scale reproduction pleasing to the eye. They use a serial of f-numbers to adapt to the brightness and at a certain f-number the dynamic range is only thousands of lux.

10.3 Evaluation of Illumination Conditions

We will now give some quantitative criteria to evaluate the quality of illumination conditions in a specified vision task. These criteria reflect the factors of signal-to-noise ratio (SNR), linearity, contrast, and natural properties of the object.

10.3.1 SNR

SNR is one of the image fidelity criteria that is an important factor in considering illumination control and is measured by determining the amount of random noise on the visual signal in an area of the scene (object). A higher number of SNR produces a picture with enhanced sharpness or other attributes.

The SNR is defined as

$$\text{SNR} = 10\log_{10} \frac{\int_{\Omega} i(x,y)^2 \, d\omega}{\int_{\Omega} [i(x,y) - \hat{i}(x,y)]^2 \, d\omega} \, [\text{dB}] \tag{10.1}$$

where $i(x, y)$ is the input signal (scene information) under certain illumination conditions, $i(x, y)$ is the corresponding noised signal, and Ω is the whole field of view.

However, noise measurement is affected by the use of aperture correction or image enhancement and may also be affected by the presence of shading or nonuniform illumination.

Noise generation in CCD imagers has several sources. The fundamental noise level results from the quantum nature of the incident light – as the light on a pixel reduces, fewer light quanta and, thus, fewer electrons are involved, and the signal gets noisier. However, this is usually not a serious limit. More important is the dark current performance of the CCD; this is a small current that flows in the absence of light input. It depends on temperature and may vary from pixel to pixel. Random fluctuations of the dark current are visible as random noise. Another CCD noise source is reset noise, which originates in the readout circuit on the chip. Furthermore, the input amplifier is also a noise source.

10.3.2 Dynamic Range

A typical CCD sensor has a limited dynamic range of illumination intensity. That is, the image irradiance l must lie in the range:

$$L_{\min} \le l \le L_{\max}. \tag{10.2}$$

For example, a "SONY XC003P - 3 CCD Color Camera" has a minimum sensitivity of 31 lux (at F2.2, GAIN +18dB, 100% level) and a normal sensitivity of 2000 lux (at F5.6). The maximum sensitivity is usually not specified because most cameras can automatically handle highlights using a knee slope and white clipping to compress the contrast.

The contrast compression knee is usually at about 90% of the *reference white* level, over which it will cause nonlinearity and loss of scene information. Once the white clip level is reached, all color will be lost and the highlight will appear white.

10.3.3 Linearity

The above interval $[L_{min}, L_{max}]$ is called the gray scale. Common practice is to shift this interval numerically to the interval $[0, L]$ by looking it up in the quantization table. $l=0$ is considered black and $l=L$ is considered white in the scale. However, the *memory look-up table* is not always linear because of transfer-characteristic processes of the camera, such as gain control, gamma correction, and highlight compression. Usually it has better linearity between 5% and 85% of total gray levels, which corresponds to 12 and 216 of 8-bit signal levels.

10.3.4 Contrast

Original contrast is important in machine vision tasks because it means obtaining clear object surface information. Although contrast can also be enhanced during post-processing (e.g. histogram equalization) of the acquired images, original contrast must be good enough so that it survives the quantization process. High original contrast may help the robot vision system to achieve a better interpretation of the scene.

Considering two surface points A and B, the survival probability of contrast (apparent difference between A and B) is

$$p_s = \min(1, \frac{|I_A - I_B|}{\Delta}),\tag{10.3}$$

where Δ is the length of the quantization step, I_A and I_B are illumination intensities of point A and B, respectively.

10.3.5 Feature Enhancement

The features of interest in machine vision include the geometrical object shape and optical surface properties, which both are represented through reflective responsity and the color vector. Therefore, another purpose of illumination control for feature enhancement is to: (1) improve the contrast of reflective responsity, (2) reflect the true color of the object surface. To achieve this purpose, we need to select the proper luminaire type and carefully control luminaire pose, radiant intensity, and color temperature.

10.4 Controllable Things

10.4.1 Brightness

The minimum light input is limited due to the dark current performance of the CCD, which depends on temperature and may vary from pixel to pixel. Most present CCDs can be sensitive from 2 to 20 (lx) at minimum. On the other hand, a brighter scene may bring higher SNR because it contains a larger signal with the same noise and higher image contrast. The basic nature of image brightness $l(x, y)$ is usually characterized by two components: illumination $i(x, y)$ and reflectance $r(x, y)$:

$$l(x, y) = i(x, y)r(x, y). \tag{10.4}$$

Under the same illumination condition, considering two surface points A and B, it is obvious that larger illumination implies a higher contrast between them because:

$$\text{Contrast} = |l_A - l_B| = i(x, y)| r_A(x, y) - r_B(x, y)|. \tag{10.5}$$

However, too bright an illumination will result in the camera's white balance clipping function and loss of object surface information (both discontinuities and colors).

10.4.2 Color Temperature and Color Rendering Index

The color in an image is derived from a complex combination of incoming illumination, material interaction, and detection parameters. The *color temperature* describes the appearance of a light source when someone looks at the light itself and the *color rendering* is given to surfaces when it shines on them. Generally, illumination in an ideal vision system should be white, which means it includes a broad spectrum of colors. But in a real environment, the light color depends on the types of light sources and their temperatures. Color temperature is the temperature of a blackbody radiator that produces a matching visual sensation to the illuminant. The color temperature of common white light sources ranges from approximately 2700 to 6500 K. For example, incandescent illumination has a color temperature in the range of 3,000 K and is seen as "reddish". Daylight usually is defined as a color temperature of 6,500 K. The locus of color temperatures shows on the CIE (http://www.cie.co.at) chromaticity diagram as a line beginning at the red end of the spectrum for low color temperatures, and curving out toward the center of the diagram for high temperatures. Video cameras have no natural ability to adapt to the illuminant. They must be told what color in the image to make "white" and then a white balancing procedure must be performed.

Practically, while the light from a lamp appears white to humans, a color CCD-camera produces a red rich image. This color variation is due to the imbalance of the lamp's spectral output, and it is further exaggerated by the wavelength-dependent sensitivity of a standard silicon CCD sensor, which has stronger sensitivity to red photons than to blue photons. In many color applications, the use of a balanced white light source is preferred in combination with an off-the-shelf single-chip color camera, with RGB output and good long-term stability, providing an optimum balance between color quality and cost.

The *color rendering index* expresses how a light source compares with natural light in its ability to make objects appear in their natural colors. It is a measure of the degree to which the colors of surfaces illuminated by a given light source conform to those of the same surfaces under a reference light. Perfect agreement is given a value of 100%. Common lamps have rendering indices ranging from 20% to 90%. For example, incandescent lamps – 90%, fluorescent tubes – 60%–90%, high-pressure mercury lamps – 40%–60%, low-pressure sodium lamps – 20%–40%.

Incandescent lamps or filament tungsten halogen lamps are stable, have a fairly long life time, good color rendering, relatively high efficiency, and are easy to install. Spectral irradiance can also be made more uniform by grinding the surface of a glass bulb. Therefore they are good options for machine vision use. A one-point light source is easy for lighting installations, but the disadvantage is non-uniform spatial distribution of illumination intensity.

10.4.3 Glare

High contrast between a luminaire and its background may produce glare. A machine eye may not be able to adapt to this situation because it exceeds the dynamic range of the cameras. It is a cause of visual discomfort because the machine eye must handle the highlight. The contrast is degraded when the highlight compression knee is reached and all color and contrast will be lost when white clip levels are reached. In this case, the robot may have difficulty to understand the scene.

Two types of glare are distinguished: (1) discomfort glare or direct glare, resulting in physical discomfort; (2) disability glare or indirect glare, resulting in a loss in visual performance. They will be discussed in the next sections.

10.4.4 Uniform Intensity

Lighting with uniform spatial distribution is the most efficient solution for a vision system. If the intensity distribution is non-uniform and the pattern is uncalibrated, it will become an additional source of noise and the SNR is degraded.

If the pattern of light source radiation is previously known (through the illumination calibration technique (CIBSE 1985)), we may obtain scene features using:

$$r(x, y) = l(x, y)/i(x, y), \tag{10.6}$$

where $r(x, y)$ reflects the optical properties (edge discontinuities and colors) of the object.

10.5 Glare Avoidance

To the human eye, glare is a source of discomfort because the high contrast between a luminaire and its background exceeds its adaptive dynamic range. This is an even worse situation for the machine eye because the vision sensor has a smaller adaptive dynamic range. Too much illumination volume will automatically cause highlight compression or white clipping. Furthermore, glare usually causes loss of the object's natural color. Hence, two types of glare, disability glare and discomfort glare, should be avoided as much as possible.

There are at least two reasons for vision system to avoid glares. First, a vision sensor has a limited dynamic range. Too much illumination volume will automatically cause highlight compression or white clipping. Second, glare usually causes loss of an object's natural color.

10.5.1 Disability Glare

Disability glare is usually caused by indirect glare and results in a loss of visual performance. Nayar et al. (1991) find that the image irradiance is a linear combination of three components, diffuse lobe I_d, specular lobe I_{s1}, and specular spike I_{s2} (Fig. 10.1):

$$I = I_d + I_{s1} + I_{s2} = k_d \cos\theta + k_{s1} e^{-\frac{\alpha^2}{2\sigma^2}} + k_{s2}\delta(\theta_c - \theta_i)\delta(\varphi_c - \varphi_i) \tag{10.7}$$

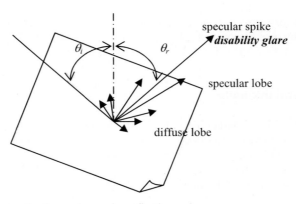

Fig. 10.1. Diffuse reflection and specular reflection

Furthermore, Gudrun et al. (1990) concluded that the light color body reflection (diffuse reflection) is determined by intrinsic characteristics of surface materials and the fact that the light color of a surface reflection (specular lobe + specular spike) has the same color as the illumination. For example, a shiny red ball will have a specular highlight on its surface only at the position where the ball's curved surface meets the normal reflection condition. The highlight has the same color as the illuminant (white) whereas, at all other positions on the ball, the reflection is diffuse and appears red.

The disability glare light causes two problems. One is that the specular reflection contains only source color, which results in the loss of color rending, causing the robot to have possible difficulty in detecting natural features of its scene. The other is that the highlight usually has a large volume of illumination intensity, which results in highlight compression or white clipping.

Practically, we can avoid the disability glare by presenting the target with light mostly from the side, so that the specularly reflected and hence brightest light is reflected off to the side and away from the field of view.

10.5.2 Discomfort Glare

In a lighting system for machine vision, although the main problem may be disability glare, in which the brightness of the luminaires may dazzle and prevent obstructions from being seen, we should also consider discomfort glare in the robot environment. It is usually caused by direct glare (due to the lighting installation) and results in physical discomfort.

There are many criteria to evaluate the glare indices. For example, the IES Technical Report "Evaluation of Discomfort Glare" (CIBSE 1985) sets out the procedure for the evaluation of the glare index in the formula:

$$\text{Glare Index} = 10\log_{10}[0.5 \times \text{constant} \sum \frac{B_s^{1.6} \omega^{0.8}}{B_b} \times \frac{1}{p^{1.6}}]. \qquad (10.8)$$

Recently, the Commission Internationale de l'Eclairage (CIE) established a new glare rating procedure known as the Unified Glare Rating system (UGR) (Einhorn 1998, Iwata and Tokura 1998) in the form of:

$$UGR = 8\log_{10}[\frac{0.25}{L_b} \sum \frac{L_s^2 \omega}{p^2}], \qquad (10.9)$$

where p is the positional index:

$$1/p = [d^2/(0.9d^2 + 2.3d + 4) - 0.1]$$
$$\times \exp(-0.17s^2/d + 0.013s^3/d) + 0.09$$
$$+ (0.075 - 0.03/d)/[1 + 3(s - 0.5)^2]$$

These criteria are initially proposed for the purpose of human visual comfort. According to the comparison of the machine eye and the human eye, the CIE-UGR criterion may be adopted for the design of a lighting system in a robot environment. A glare index below UGR-19 is acceptable and above UGR-25 is uncomfortable.

If the light source itself remains in the field of view and is bright, it can become a source of discomfort glare. Therefore it is best to position the light source behind the camera, either above or to the side. We can also reduce the effects of discomfort glare by increasing the task luminance relative to the luminance of the surroundings. Discomfort glare can also be reduced by: (1) decreasing the luminance of the light source, (2) diminishing the area of the light source, and (3) increasing the background luminance around the source if we can stop down the sensor aperture in this case.

10.6 Intensity Estimation

To satisfy the visual comfort of machine eyes, apart from selecting proper types of light sources and cameras, the key controllable parameters of a luminare are radiant intensity and geometrical pose in a practical vision system. The purpose of intensity control is to achieve proper image brightness which is in the range of the sensor, with linear property, and has contrast as high as possible. The purpose of pose control is to avoid possible glare and achieve uniform intensity distribution.

To control the image intensity so that it will concentrate on an optimal point, firstly the sensor sensitivity must be considered, then the image irradiance is estimated from source radiation to image sensing, and finally the optimal control point is decided.

10.6.1 Sensor Sensitivity

The brightness that the camera perceived is the intensity (energy level) of light and varies depending on the light color (wavelength). The sensors usually have most sensitivity at the wavelength of 555 nm with the corresponding efficiency defined as 100%. The distribution is quantified by a curve of brightness sensation versus wavelength, called the luminosity curve. Figure 10.2a illustrates the sensor quantum efficiencies of back-illuminated CCD and front-illuminated CCD (Gilblom 1998). Since a video camera must have a spectral response that matches the human luminosity curve, the curves of sensor sensitivity have been normalized so their areas are equal, illustrated in Fig. 10.2b for 1-CCD and in Fig. 10.2c for 3-CCD. The sensor sensitivity curves are expressed as:

$$\rho_r = \rho_r(\lambda), \rho_g = \rho_g(\lambda), \rho_b = \rho_b(\lambda). \tag{10.10}$$

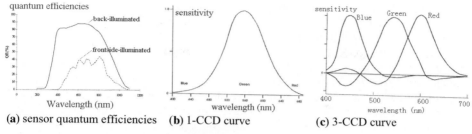

(a) sensor quantum efficiencies   **(b)** 1-CCD curve   **(c)** 3-CCD curve

Fig. 10.2. Sensor sensitivity (Gilblom 1998)

10.6.2 Estimation of Image Irradiance

To estimate the image irradiance, we need to analyze five procedures, i.e. source radiation, source efficiency, surface irradiance, surface reflection, and sensor perception. First, the total output radiation of a light source at temperature T is proportional to four times the temperature:

$$M_e = \varepsilon\sigma T^4 \tag{10.11}$$

where $\sigma = 5.67051 \times 10^{-8} \mathrm{Wm^{-2}K^{-4}}$ and the emissivity $\varepsilon \in [0, 1]$ varies with wavelength.

According to *Planck's radiation law*, the spectral distribution of the radiation emitted by a blackbody can be described as a function of wavelength λ,

$$M(\lambda) = \frac{C_1}{\lambda^5}\frac{1}{e^{C_2/\lambda T} - 1} \tag{10.12}$$

where C_1 and C_2 are two radiation constants,

$$C_1 = 2\pi hc^2 = 3.741774 \times 10^{-16}\ \mathrm{Wm^2}$$
$$C_2 = \frac{hc}{k} = 0.01438769\ \mathrm{mK}.$$

The peak wavelength, $\lambda_{\max}$, in nanometers, is given by

$$\lambda_{\max} = \frac{2.8978 \times 10^6}{T} = 2.8978 \times 10^6 \left(\frac{\varepsilon\sigma}{i^2 R}\right)^{\frac{1}{4}}. \tag{10.13}$$

Equation (10.13) depicts the power output of the light source and the spectral distribution of intensity at different temperatures. As from Fig. 10.3, obviously we can find that with increasing temperature, more energy is emitted and the peak emission shifts toward the shorter wavelengths.

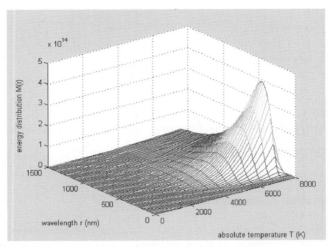

Fig. 10.3. Energy density vs spectral distribution and temperature

Figure 10.4 illustrates the energy distribution of a 100 W incandescent lamp. However, consider that the vision sensor is sensitive only to a portion of the electro-magnetic wave, i.e. $380 < \lambda < 750$ (nm). Due to the quantum efficiency of the vision sensor, the quantity of light as seen by the camera (illustrated as the grey area in Fig. 10.5.), is

$$W_e = \int_{\lambda_1}^{\lambda_2} M(\lambda) \cdot \rho(\lambda) \mathrm{d}\lambda = \int_{350}^{750} \frac{C_1}{\lambda^5} \frac{1}{e^{C_2/\lambda T} - 1} \rho(\lambda) \mathrm{d}\lambda, \qquad (10.14)$$

where W_e is the efficient light energy.

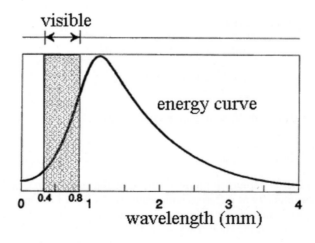

Fig. 10.4. Distribution of source radiation

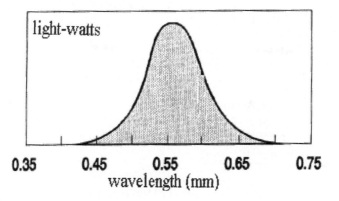

Fig. 10.5. Efficiency of light energy

To control the quantity of source radiation, a dimmer is usually employed to adjust the phase of AC waveforms. Denote the input power rate $\eta(\phi)$ and efficiency $\varepsilon(\phi)$ at phase ϕ controlled by the dimmer, then the visible efficient energy is:

$$W_e = \int_{\lambda_1}^{\lambda_2} M(\lambda) \cdot \rho(\lambda) \cdot \eta(\phi) \cdot \varepsilon(\phi) \mathrm{d}\lambda. \tag{10.15}$$

In the case of a 3-CCD camera, since the color temperature of the light source varies as long as the input power changes, the visible efficient energy becomes:

$$W_{r,g,b} = \int_{\lambda_1}^{\lambda_2} M(\lambda) \cdot \rho_{r,g,b}(\lambda) \cdot \eta(\phi) \cdot \varepsilon_{r,g,b}(\phi) \mathrm{d}\lambda. \tag{10.16}$$

On the other hand, we have

$$L = \int_{380}^{750} \frac{C_1}{\lambda^5} \frac{1}{e^{C_2/\lambda T} - 1} \mathrm{d}\lambda \underset{x=\lambda T}{=\!=\!=} T^4 \int_{380T}^{750T} \frac{C_1}{x^5} \frac{1}{e^{C_2/x} - 1} \mathrm{d}x = C(T)T^4 \tag{10.17}$$

where

$$C(T) = \int_{380T}^{750T} \frac{C_1}{x^5} \frac{1}{e^{C_2/x} - 1} \mathrm{d}x = \lim_{\Delta x \to 0} \sum_{380T}^{750T} \frac{C_1}{x^5} \frac{1}{e^{C_2/x} - 1} \Delta x \tag{10.18}$$

is defined as a coefficient function.

The following algorithm is developed using numerical computation to solve (10.18), an example curve is illustrated in Fig. 10.6. Finally the luminous flux function (10.17) can be obtained.

```
                    "Algorithm for the solution of C(T)"
C1=3.74174*(10^(-16));
C2=0.01438769;
e=2.718281828459;
func=zeros(1000,1);

for jt=1:1000;
        T=200+jt*30;
        r1=380.0*T*(10^(-9));
        r2=750.0*T*(10^(-9));
        temp=0.0;
        for i=1:3000;
            r=r1+i*(r2-r1)/3000.0;
            temp=temp+((r2-r1)/3000.0)*C1/(((r^5)*((e^(C2/r))-1));
        end;
        func(jt)=temp;
end;
```

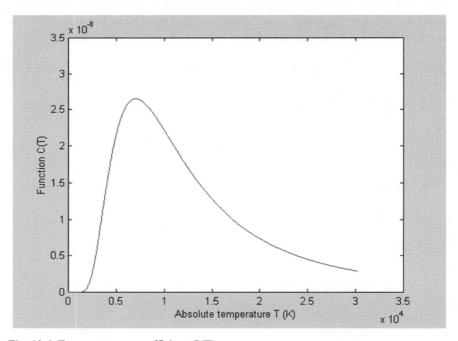

Fig. 10.6. Temperature vs coefficient $C(T)$

In most cases, the light emitted by a source is usually not uniformly distributed in all directions and the luminous intensity varies according to the position beneath the source. Manufacturers of luminaires usually provide intensity distribution diagrams for their products, which show the relationship between the luminous intensity and the angle to a reference position of 0° situated vertically below the source. The polar graph is often used for these purposes. It can also be calibrated using the techniques of luminaire photometry developed by Lewin and John (1999). Finally the efficient energy distribution is modeled as:

$$L(\phi,\theta) = W_{r,g,b} \cdot \Gamma(\phi,\theta), \tag{10.19}$$

where the function $0 \leq \Gamma(\phi, \theta) < 1$ describes the spatial distribution of source radiation.

Considering a point on the object surface, its irradiance is the integral of the whole angular distribution over a specified solid angle, i.e.

$$L_s = \int_0^{2\pi} \int_0^{\pi/2} L(\phi,\theta) \cos\phi \sin\phi \cdot d_\phi d_\theta. \tag{10.20}$$

The object surface then becomes another source and the image irradiance of the vision sensor can also be computed

$$E = \frac{\pi}{4} (\frac{d}{f})^2 L_s \cos^4 \alpha. \tag{10.21}$$

Since real objects are usually not Lambertians, three parts contribute to the surface reflection, that is I_d (diffuse reflection), I_{s1} (gross specular reflection), and I_{s2} (specular reflection). Then the image irradiance of an object illuminated by a source is represented by a function as in, (Nayar 1991, Laszlo 1999)

$$I = \begin{bmatrix} I_r \\ I_g \\ I_b \end{bmatrix} = k'_d \cos\theta \begin{bmatrix} r_d \\ g_d \\ b_d \end{bmatrix} + k'_{s1} e^{-\frac{\alpha^2}{2\sigma^2}} \begin{bmatrix} r_{s1} \\ g_{s1} \\ b_{s1} \end{bmatrix} + k'_{s2} \delta(\theta_c - \theta_i)\delta(\varphi_c - \varphi_i) \begin{bmatrix} r_{s2} \\ g_{s2} \\ b_{s2} \end{bmatrix} \tag{10.22}$$

where $\delta(x)$ is unit pulse function or Dirac delta function.

10.7 Intensity Control

10.7.1 The Setpoint

A camera usually has the requirement of minimum illumination which is typically 2 [lux] with high-gain operation. Theoretically, a camera's sensitivity could be increased as much as desired simply by increasing the amplifier gain and operating the CCDs at a lower output level. Of course, the SNR will degrade when this is done. That is what happens in a camera's "high gain" modes, which trade signal quality for sensitivity. On the other hand, the full-quality mode of a camera operates the CCDs at the light level given in the sensitivity specification. This may be somewhat of a trade-off with highlight performance.

The vision sensor often has best linearity between 15% and 90% of the output level (Fig. 10.7). In fact, the illumination condition below 20% is unacceptable because: (1) low SNR for the existence of noise and dark current, (2) nonlinear quantization at this area, (3) nonlinearity because of gamma correction. An illumination condition above 90% output level is also unacceptable because of contrast compression of the knee slope and loss of color properties. Hence, the optimal setpoint of illumination intensity is at about 80% of the output level because of high SNR, linearity, and contrast.

The illumination intensity can be controlled in two ways: (1) phase-control to adjust the electrical current intensity using a dimmer; (2) pose-control to adjust the distance between object and luminaire using a robot end-effector. Usually it is better to keep the luminaires far away from the object because the illumination will be more uniform in this case and will increase image SNR. It is also better to keep the luminaire in full-on state because it entails a higher color rending index in this condition and facilitates the obtaining of true surface information.

Machine vision applications have a great need for feedback control. The vision-illumination system can be considered a closed-loop system in which the vision sensor plays a second role as the feedback channel. The pose and dimmer phase are determined by a controller according to the visual feedback, source model, and optimal setpoint. The energy magnitude of source radiation and image irradiance may be estimated using the techniques discussed above.

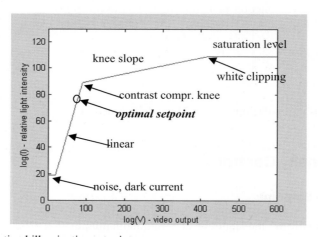

Fig. 10.7. Optimal illumination setpoint

10.7.2 System Design

Figure 10.8 illustrates the overview of a typical system for illumination control, which includes a robot, manipulators of the light source and vision sensors, an image processor, a system controller, and an object in the scene.

Here we focus on the control of the energy magnitude of source radiation, although other parameters may be discussed in future. The block diagram of the illumination control system is illustrated in Fig. 10.9, where the symbols mean:

– x_e: the disturbance of the environment. It results from three reasons: the changing natural light, the dynamic environment, and the moving vision sensor;
– x_o: the sensed image (the output of vision preprocessing).
– x_r: the image irradiance after the displacement of the vision sensor and the illuminants.
– s_{si}: the setpoint of parameters, the output value for setting the vision sensor's parameters and illuminant's parameters. It is a vector $s_{si}=[s_s, s_i]^T$.

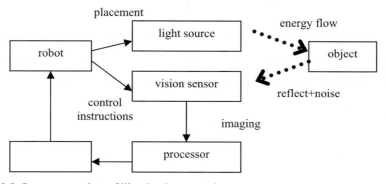

Fig. 10.8. System overview of illumination control

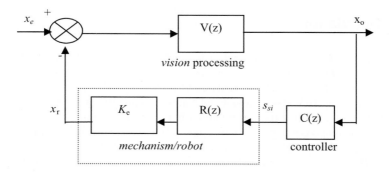

Fig. 10.9. Block diagram of illumination control

- K_e: the gain which relates the image irradiance and the parameters of both vision sensor and illuminant (geometrical pose and optical settings).
- $V(z)$: the vision system. Generally, $V(z)=k_v/z^n$, which means the vision system introduces n unit delays.
- $C(z)$: the controller. It can be a fuzzy-controller or another robust and intelligent controller. It satisfies all above-mentioned constraints and keeps the image from becoming too dark or too bright. The control parameters are based on image irradiance distribution (a statistical value or vector based on the image histogram).

It is an active system with visual feedback. The goal is to keep the object sensed in a good illumination condition to provide good results for further visual processes. In our simulation system, only for testing the principle, a fuzzy-PID controller is used to adjust the parameters of the illuminant. The closed-loop transfer function is:

$$H(z) = \frac{x_o(z)}{x_e(z)} = \frac{V(z)}{1 + K_e \cdot C(z) \cdot R(z) \cdot V(z)}.$$ (10.23)

where $V(z) = \dfrac{k_v}{z^n}$ means that the vision system introduces n unit delays for

acquiring an image and processing the data. Assume the mechanism (integrator, inverse kinematics, and servo, etc) also produce m unit delays, that is $R(z) = \dfrac{1}{z^m}$.

The transfer function therefore becomes

$$H(z) = \frac{\dfrac{k_v}{z^n}}{1 + K_e \cdot C(z) \cdot \dfrac{1}{z^m} \cdot \dfrac{k_v}{z^n}} = \frac{k_v z^m}{z^{m+n} + K_e \cdot k_v \cdot C(z)}.$$ (10.24)

10.8 Simulation

A simulation system for illumination control was implemented with MATLAB (Fig. 10.10). The step response, sine response, zero input response, and random input response of the actively illuminated vision system have been observed, while we assume it has 10% environment light noise. Typically, step response happens when the robot stays in a dark room and a light is turned on at a certain time (Fig. 10.11). Sine response happens when the robot walks in an environment with periodically installed lights (Fig. 10.12), for example, a robot moving on a road as in Fig. 10.13. Random input response happens usually in a general natural environment (Fig. 10.14).

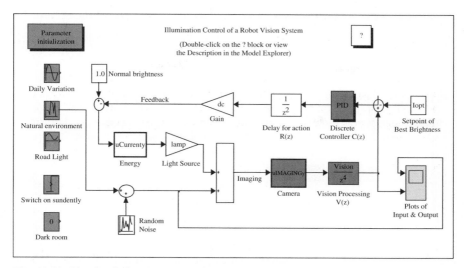

Fig. 10.10. The simulation system for illumination control

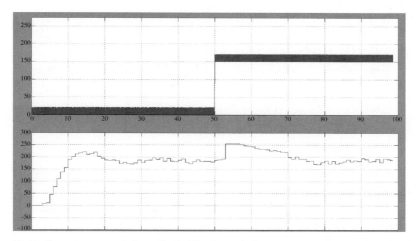

Fig. 10.11. Step response of the actively illuminated vision system, with 10% environment light noise. (It happens when the robot stays in a dark room and a light is turned on at a certain time)

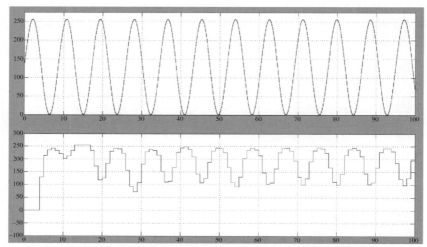

Fig. 10.12. Sine response of the actively illuminated vision system. (It happens when the robot walks in an environment with periodically installed lights, e.g. on a road as in Fig. 10.13)

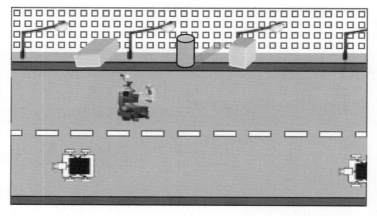

Fig. 10.13. A robot walking in a virtual environment with periodically installed lights

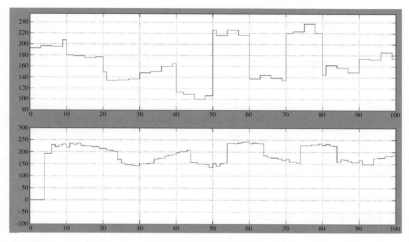

Fig. 10.14. Response of the actively illuminated vision system in a natural random environment, with 10% noise. (In a general environment)

10.9 Implementation

10.9.1 Design for Active Illumination

With thermal light sources, such as electric light, we consider its input power M_0:

$$M_0 = u \cdot i = i^2 R \text{ (watt/s)} \tag{10.25}$$

Since $M_e = \varepsilon \sigma T^4$ (10.11), we have

$$\varepsilon \sigma T^4 = i^2 R \text{ or } T = \left(\frac{i^2 R}{\varepsilon \sigma}\right)^{\frac{1}{4}} = \alpha \sqrt{i} \tag{10.26}$$

where the coefficient $\alpha = \left(\dfrac{R}{\varepsilon \sigma}\right)^{\frac{1}{4}}$ is a constant.

In this way, the radiant intensity and source temperature can be adjusted using a dimmer for phase control. The simplest way is to generate a PWM signal and compare the message signal to a triangular or ramp waveform (Fig. 10.15). The hardware of control module in our laboratory includes a controller board. The controller based on the Atmega16L processor provides several sensor channels, 8.4-24 V input, serial ports (TTL and RS232), PWM mode outputs and on-off mode outputs. As can be seen, we built this module for lighting control. The generated luminance is further calibrated once for generating a control curve.

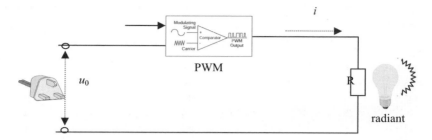

Fig. 10.15. A simple implementation for digital control of radiant intensity and spectral distribution

10.9.2 Experimental Robots

Currently, we are also working on setting up a robot system with active illumination control for testing in real environments (Fig. 10.16) at the University of Hamburg.

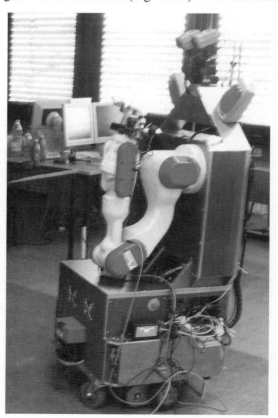

Fig. 10.16. The mobile robot setup in our laboratory for dexterous manipulation

The mobile robot has an end-effector with 6DOFs for dexterous manipulation. An eye-in-hand camera is used to observe the scene and the active illumination device contains two light sources which can be controlled by a PWM module (Fig. 10.17). The robot works in many different environments and controls its illumination level automatically. Figure 10.18 illustrates an example of active sensing where the red curve is the brightness of the input image and the blue curve is the output of the lighting level (in percentage). It results in the image sequence being kept in a good range of brightness for scene understanding. A next step of our work is to equip a mobile robot with controllable illumination for active search and recognition tasks, which cannot only adjust the illumination level but also change the lighting direction and avoid glares (Fig. 10.19).

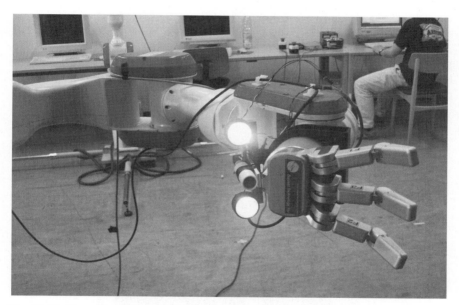

Fig. 10.17. The active illumination device (one eye-in-hand camera and two light sources controlled by a PWM module)

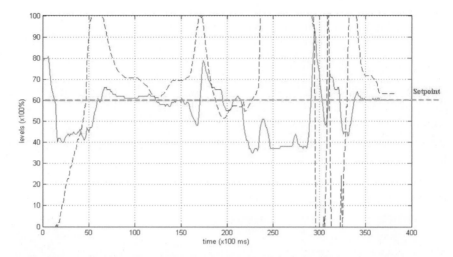

Fig. 10.18. An example of illumination control for active sensing where the solid curve is the brightness of the input image and the dashed curve is the output of the lighting level (in percentage)

Fig. 10.19. A robot equipped with controllable illumination for autonomous tasks of active search and recognition

10.10 Summary

This chapter presented an idea of active illumination control for robot vision. Strategies are proposed to achieve optimal illumination conditions for vision sensors so that best-quality images can be obtained with high SNR, contrast, color rending, and linearity. The controllable parameters include optical parameters and pose parameters of luminaire and sensor. The characteristics of a robot eye and its "comfort" conditions have been analyzed. The image intensity is theoretically controlled at a good setpoint. Glare avoidance methods are proposed for treating two types of glare, disability glare and discomfort glare. The disability glare can be eliminated by placements of the light source, vision sensor, or targets. The discomfort glare is evaluated using the CIE-UGR criterion and can be diminished by control of the source position, radiant flux, or background luminance.

10.10 Summary

Bibliography

A

B. A. Abidi, "Automatic sensor placement". *Proceedings of SPIE Conference on Intelligent Robots and Computer Vision XIV*, Philadelphia, October (1995) Vol. 2588, pp. 387–398.

S. Abrams, P. K. Allen, and K. Tarabanis, "Computing camera viewpoints in an active robot work cell", *International Journal of Robotics Research*, Vol. 18, No. 3, pp. 267–288, March 1999.

J. Adachi and J. Sato, "Uncalibrated visual servoing from projective reconstruction of control values", *2004 IEEE Computer Society Conference on Computer Vision and Pattern Recognition*, Washington, DC, USA, 2004, pp. 297–300.

Y. Adachi, H. Tsunenari, Y. Matsumoto, and T. Ogasawara, "Guide robot's navigation based on attention estimation using gaze information", *2004 IEEE/RSJ International Conference on Intelligent Robots and Systems*, Sendai, Japan, 2004, pp. 540–545.

W. B. Agocs, "Least squares residual anomaly determination", *Geophysics*, Vol. 16, pp. 686–696, 1951.

N. Ahuja and J. Veenstra, "Generating octrees from object silhouettes in orthographic views", *IEEE Transaction on Pattern Analysis and Machine Intelligence*, Vol. 11, No. 2, pp. 137–149 , February 1989.

G. Aggarwal and R. Chellappa, "Face recognition in the presence of multiple illumination sources", *Tenth IEEE International Conference on Computer Vision*, Beijing, China, 2005, pp. 1169–1176.

M. Z. Alan, K. Lindsay, and R. A. Russell, "Topological mapping inspired by techniques in DNA sequence alignment", *IEEE/RSJ International Conference on Intelligent Robots and Systems*, Beijing, China, 2006, pp. 2754–2759.

T. Aldinger and J. Kao, "Data fusion and theater undersea warfare – an oceanographer's perspective", *Proceedings of MTS/IEEE OCEANS*, Washington, DC, USA, 2004, pp. 2008–2012.

A. Alempijevic and G. Dissanayake, "An efficient algorithm for line extraction from laser scans", *2004 IEEE Conference on Robotics, Automation and Mechatronics*, Singapore, 2004, pp. 970–974.

C. R. Allen and I. C. Leggett, "3D scene reconstruction and object recognition for use with autonomously guided vehicles (AGVs)", *IEEE IECON*, 1995, pp. 1219–1224.

Y. Aloimonos, "Introduction: Active vision revisited", *Active perception*, Hillsdale, N. J. Lawrence Erlbaum Associates, 1993, pp. 1–18.

M. Ammi and A. Ferreira, "Haptically generated paths of an AFM-based nanomanipulator using potential fields", *4th IEEE Conference on Nanotechnology*, Bourges, France, 2004, pp. 355–357.

U. Anliker, J. Beutel, M. Dyer, R. Enzler, P. Lukowicz, L. Thiele and G. Troster, "A systematic approach to the design of distributed wearable systems", *Transaction on Computers*, Vol. 53, No. 8, pp. 1017–1033, 2004.

G. Antonelli and S. Chiaverini, "Experiments of fuzzy lane following for mobile robots", 2 ed 2004, pp. 1079–1084.

M. Antonio, et al., "Integrated grasp planning and visual object localization for a humanoid robot with five-fingered hands", *IEEE/RSJ International Conference on Intelligent Robots and Systems*, Beijing, China, October 2006, pp. 5663–5668.

T. Arbel and F. P. Ferrie, "Viewpoint selection by navigation through entropy maps", *Seventh IEEE International Conference on Computer Vision*, Vol. 1, 1999, pp. 248–254.

T. Arbel and P. Ferrie, "Entropy-based gaze planning", IEEE *Transactions on Systems, Man, and Cybernetics-Part B: Cybernetics*, Vol. 19, No. 11, pp. 779–786, September 2001.

M. Ashdown, M. Flagg, R. Sukthankar, and J. M. Rehg, "A flexible projector-camera system for multi-planar displays", *2004 IEEE Computer Society Conference on Computer Vision and Pattern Recognition*, Washington, DC, USA, 2004, p. II-165.

H. Ashraf, S. Peter, and D. Rudiger, "Autonomous feature-based exploration using multi-sensors", *IEEE International Conference on Multisensor Fusion and Integration for Intelligent Systems*, Heidelberg, Germany, 2006, pp. 456–461.

B

R. Bajcsy, "Active perception", *Proceedings of the IEEE*, Vol. 76, No. 8, August 1988, pp. 996–1005.

M. Bajracharya, A. az-Calderon, M. Robinson, and M. Powell, "Target tracking, approach, and camera handoff for automated instrument placement," *2005 IEEE Aerospace Conference*, Pasadena, USA, March 2005, pp. 52–59.

J. E. Banta and M. A. Abidi, "Autonomous placement of a range sensor for acquisition of optimal 3-D models", *22nd IEEE International Conference on Industrial Electronics, Control, and Instrumentation*, Vol. 3, pp. 1583–1588, August 1996.

N. Bart, H. Derek, A. A. E. Alexei, and H. Martial, "Opportunistic use of vision to push back the path-planning horizon", *IEEE/RSJ International Conference on Intelligent Robots and Systems*, Beijing, China, October 2006, pp. 2388–2393.

A. Bartoli, P. Sturm, and R. Horaud, "Structure and motion from two uncalibrated views using points on planes", *Proceedings of the Third International Conference on 3D Digital Imaging and Modeling*, Quebec City, Canada, June 2001, pp. 83–90.

B. G. Batchelor and P. F. Whelan, "Machine vision systems: Proverbs, principles, prejudices and priorities", *Proceedings of the SPIE on Machine Vision Applications, Architectures, and Systems Integration III*, Boston, USA, 1994, Vol. 2347, pp. 374–383.

J. Batlle, E. Mouaddib, and J. Salvi, "Recent progress in coded structured light as a technique to solve the correspondence problem: A survey", *Pattern Recognition*, Vol. 31, No. 7, pp. 963–982, July 1998.

A. E. bdel-Hakim and A. A. Farag, "Robust virtual forces-based camera positioning using a fusion of image content and intrinsic parameters", *8th International Conference on Information Fusion*, 2005, p. 8.

K. E. Bekris, A. A. Argyros, and L. E. Kavraki, "Angle-based methods for mobile robot navigation: Reaching the entire plane", *2004 IEEE International Conference on Robotics and Automation*, Barcelona, Spain, 2004, pp. 2373–2378.

F. Belkhouche and B. Belkhouche, "Modeling and controlling a robotic convoy using guidance laws strategies", *Systems, Man and Cybernetics, Part B, IEEE Transactions on*, Vol. 35, No. 4, pp. 813–825, 2005.

P. J. Besl and N. D. Mckay, "A method for registration of 3-D shapes", *IEEE Transactions on Pattern Analysis and Machine Intelligence*, Vol. 14, No. 2, pp. 239–256, February 1992.

A. Bevilacqua, L. Di Stefano, and P. Azzari, "People tracking using a time-of-flight depth sensor", *IEEE International Conference on Video and Signal Based Surveillance*, Washington, DC, USA, 2006, p. 89.

S. Bhattacharya, R. Murrieta-Cid, and S. Hutchinson, "Optimal paths for landmark-based navigation by differential-drive vehicles with field-of-view constraints,", *IEEE Transactions on Robotics*, Vol. 23, No. 1, pp. 47–59, 2007.

G. M. Bianco, "Getting inspired from bees to perform large scale visual precise navigation", 1 ed 2004, pp. 619–624.

H. Bingwei and Y. F. Li, "A Self-termination judgment method in 3D object automatic measurement and inspection", *Sixth World Congress on Intelligent Control and Automation,* Dalian, China, 2006, pp. 5008–5012.

L. Birnbaum, M. Brand, and P. Cooper, "Looking for trouble: Using causal semantics to direct focus of attention", *ICCV-93,* pp. 49–56, 1993.

M. Bjorkman and D. Kragic, "Combination of foveal and peripheral vision for object recognition and pose estimation", *2004 IEEE International Conference on Robotics and Automation,* Barcelona, Spain, 2004, pp. 5135–5140.

P. Blaer and P. K. Allen, "View planning for automated site modeling, *IEEE International Conference on Robotics and Automation,* May 2006, Orlando, pp. 2621–2626.

F. Blais, "Review of 20 years of range sensor development" *Journal of Electronic Imaging,* Vol. 13, No. 1, pp. 231–240, 2004.

N. Blanc, T. Oggier, et al., "Miniaturized smart cameras for 3D-imaging in real-time [mobile robot applications]", *The 3rd IEEE Conference on Sensors,* Vienna, Austria, October 2004, pp. 471–474.

R. Bodor, A. Drenner, et al., "Mobile camera positioning to optimize the observability of human activity recognition tasks", *IEEE/RSJ International Conference Intelligent Robots and Systems,* Edmonton, Canada, August 2005a, pp. 1564–1569.

R. Bodor, P. Schrater, and N. Papanikolopoulos, "Multi-camera positioning to optimize task observability", *IEEE Conference on Advanced Video and Signal Based Surveillance,* Como, Italy, 2005b, pp. 552–557.

R. Bolter and F. Leberl, "Shape-from-shadow building reconstruction from multiple view SAR images", *24th Workshop of the Austrian Association for Pattern Recognition,* Villach, Carinthia, Austria, Band 142, pp. 199–206, 2000.

H. Borotschnig and L. Paletta, "Appearance-based active object recognition", *Image and Vision Computing,* Vol. 18, No. 9, pp. 715–727, June 2000.

U. Borzenko, X. Wei, et al., "Lights and camera: Intelligently controlled multi-channel pose estimation system", *International Conference on Computer Vision Systems,* New York, 2006, p. 42.

L. Brinbaum, et al., "Looking for trouble: Using casual semantics to direct focus of attention", *International Conference on Computer vision,* Berlin, Germany, May 1993, pp. 49–56.

C

C. Caccavale, V. Lippiello, et al., "RePLiCS: An environment for open real-time control of a dual-arm industrial robotic cell based on RTAI-Linux", *2005 IEEE/RJS International Conference on Intelligent Robots and Systems,* Edmonton, Canada, 2005, pp. 2493–2498.

R. L. Canceroni and K. N. Kutulakos, "Toward recovering shape and motion of 3D curves from multi-view image sequences", *IEEE Conference on Computer Vision and Pattern Recognition,* Fort Collins, Colorado, 1999, pp. 192–199.

D. Chapuis, R. Gassert, L. Sache, E. Burdet, and H. Bleuler, "Design of a simple MRI/fMRI compatible force/torque sensor", *IEEE/RSJ International Conference on Intelligent Robots and Systems,* Sendai, Japan, 2004, pp. 2593–2599.

V. Chaudhary, "Surgery needs HPC, Really!", *2006 International Conference Workshops on Parallel Processing,* Columbus, USA, 2006, p. 1.

T. A. Cheema, I. M. Qureshi, A. Jalil, and A. Naveed, "Artificial neural networks for blur identification and restoration of nonlinearly degraded images", *International Journal of Neural Systems,* Vol. 11, No. 5, pp. 455–461, October 2001.

C. H. Chen and A. C. Kak, "Modeling and calibration of a structured light scanner for 3-D robot vision", *IEEE Conference Robotics and Automation,* Raleigh, USA, 1987, pp. 807–815.

L. Chen and G. Lin, "A vision-aided reverse engineering approach to reconstructing free-form surfaces", *Robotics and Computer-integrated Manufacturing*, Vol. 13, No. 4, pp. 323–336, 1997.

J. Chen, W. E. Dixon, D. M. Dawson, and V. K. Chitrakaran, "Visual servo tracking control of a wheeled mobile robot with a monocular fixed camera", 2 ed 2004, pp. 1061–1066.

J. Chen, D. M. Dawson, W. E. Dixon, and V. K. Chitrakaran, "Navigation function based visual servo control", *Proceedings of the IEEE American Control Conference,* Portland, Oregon, 2005, pp. 3682–3687.

S. Chen, R.-L. Hsu, and A. K. Jain, "Registration of range views using spherical harmonics", Technical Report, MSU-CSE-99-37, Michigan State University, November 1999.

S. Y. Chen and Y. F. Li, "Automatic sensor placement for model-based robot vision", *IEEE Transactions on Systems, Man and Cybernetics, Part B,* Vol. 34, No. 1, pp. 393–408, 2004a.

S. Y. Chen and Y. F. Li, "Active viewpoint planning for model construction", *IEEE International Conference on Robotics and Automation,* Taiwan, 2004b, pp. 4411–4416.

S. Y. Chen, W. L. Wang, G. Xiao, C. Y. Ya, and Y. F. Li, "Robot perception planning for industrial inspection", *IEEE Region 10 Conference (TENCON),* Thailand, 2004, pp. 613–616.

S. Y. Chen and Y. F. Li, "Vision sensor planning for 3-D model acquisition", IEEE Transactions on *Systems, Man and Cybernetics, Part B,* Vol. 35, No. 5, pp. 894–904, 2005.

S. Y. Chen, J. Zhang, H. Zhang, W. Wang, and Y. F. Li, "Active illumination for robot vision", *2007 IEEE International Conference on Robotics and Automation,* Rome, Italy, 2007, pp. 411–416.

D. Chengjun, C. Genqun, Z. Minglu, and D. Ping, "Mobile robot's road following based on color vision and FGA control strategy", 1 ed 2006, pp. 3124–3128.

V. Cherkassky, X. Shao, F. Mulier, and V. Vapnik, "Model complexity control for regression using VC generalization bounds", *IEEE Transactions Neural Networks,* Vol. 10, No. 5, pp. 1075–1089, 1999.

W. Chio and T. Kurfess, "Uncertainty of extreme fit evaluation for three dimensional measurement data analysis," *Computer-Aided Design*, Vol. 30, No. 7, pp. 549–557, 1995.

S. Chitale and W. T. Padgett, "Blur identification and correction for a given imaging system", *Proceedings of the IEEE Southeastcon Conference,* Tampa, USA, 1999, pp. 268–273, 1999.

A. Chivilo, F. Mezzaro, A. Sgorbissa, and R. Zaccaria, "Follow-the-leader behaviour through optical flow minimization", *IEEE/RSJ International Conference on Intelligent Robots and Systems,* Sendai, Japan, 2004, pp. 3182–3187.

V. Christoffer, L. Achim, and D. Tom, "Incremental topological mapping using omnidirectional vision," *IEEE/RSJ International Conference on Intelligent Robots and Systems,* Beijing, China, 2006, pp. 3441–3447.

C. W. Chu, S. Hwang, and S. K. Jung, "Calibration-free approach to 3D reconstruction using light stripe projections on a cube frame", *IEEE 3rd International Conference on 3D Digital Imaging and Modeling,* Quebec City, Canada, June 2001, pp. 13–19.

K. Chung-Hsien, Y. Chun-Ming, and Y. Fang-Chung, "Development of intelligent vision fusion based autonomous soccer robot", *IEEE International Conference on Hands-on Intelligent Mechatronics and Automation,* Taipei, Taiwan, 2005, pp. 124–129.

CIBSE, "TM10 – The calculation of glare indices", Chartered Institution of Building Services Engineers, London, 1985.

C. I. Connolly, "The determination of next best views", *Proceedings of the IEEE International Conference on Robotics and Automation*, St. Louis, USA, March (1985), 432–435.

T. F. Cootes, E.C. Di Mauro, et al., "Flexible 3D models from uncalibrated cameras", *Image and Vision Computing*, Vol. 14, No. 8, pp. 581–587, August 1996.

C. K. Cowan and P. D. Kovesi, "Automatic sensor placement from vision task requirements", *IEEE Transactions on pattern analysis and machine intelligence*, Vol. 10, No. 3, pp. 407–416, May 1988.

C. K. Cowan and A. Bergman, "Determining the camera and light source location for a visual task", *1989 IEEE International Conference on Robotics and Automation*, May 1989, Vol.1 pp. 509–514.

E. L. Creed, et al., "LEO-15 Observatory - the next generation", *Proceedings of MTS/IEEE OCEANS Conference,* Washington, DC, 2005, pp. 657–661.

D, E

M. Daum and G. Dudek, "On 3-D surface reconstruction using shape from shadows", *Proceedings of the 3rd 1998 IEEE Computer Society Conference on Computer Vision and Pattern Recognition,* pp. 461–468, 1998.

J. Davis and X. Chen, "A laser range scanner designed for minimum calibration complexity", *Proceedings of the 3rd International Conference on 3-D Digital Imaging and Modeling*, Ottawa, Canada, pp. 91–98, May 2001.

F. Deinzer, J. Denzler, and H. Niemann, "Viewpoint selection – a classifier independent learning approach", *Proceedings of the 4th IEEE Southwest Symposium Image Analysis and Interpretation,* pp. 209–213, 2000.

F. W. DePiero and M. M. Trivedi, "3-D computer vision using structured light: Design, calibration and implementation issues", *Advances in Computers,* Vol. 43, pp. 243–278, 1996.

F. Devernay and O. Faugeras, "From projective to Euclidean reconstruction", *Proceedings of the Computer Vision and Pattern Recognition Conference*, San Francisco, CA, pp. 264–269, June 1996.

C. D'Souza, B. H. Kim, and R. Voyles, "Morphing bus: A rapid deployment computing architecture for high performance, resource-constrained robots", *IEEE International Conference on Robotics and Automation,* Rome, Italy, 2007, pp. 311–316.

R. G. Deen and J. J. Lorre, "Seeing in three dimensions: Correlation and triangulation of Mars Exploration Rover imagery", *IEEE International Conference on Systems, Man and Cybernetics,* Hawaii, USA, October 2005, pp. 911–916.

L. Deng, F. Janabi-Sharifi, and W. J. Wilson, "Hybrid motion control and planning strategies for visual servoing", *Industrial Electronics, IEEE Transactions on*, Vol. 52, No. 4, pp. 1024–1040, 2005.

J. Dias, A. de Almeida, H. Araújo, and J. Batista, "Camera Recalibration with hand-eye robotic system", *1991 IEEE Industrial Electronics Society IECON 91*, Kobe, Japan, October 1991.

J. F. Dillenburg, P. C. Nelson, O. Wolfson, O. Yu, A. P. Sistla, S. McNeil, A. M. Ouksel, B. Xu, and J. Ben-Arie, "Applications of a transportation information architecture", *IEEE International Conference on Networking, Sensing and Control,* Taipei, Taiwan, 2004, pp. 480–485.

X. Ding, D. Piao, and Q. Zhu, "Optical imaging array design with multiple sources and detectors", *Proceedings of the IEEE 26th Annual Northeast Bioengineering Conference*, 2000, pp. 69–70.

P. M. Djuric, "Asymptotic MAP criteria for model selection", *IEEE Transactions on Signal Processing*, Vol. 46, No. 10, pp. 2726–2734, 1998.

M. Dominguez-Matas, F. J. Sainchez-Femaindez, R. Carmona-Galan, and E. Roca-Moreno, "Experiments on global and local adaptation to illumination conditions based on focal-plane average compuutation", *10th International Workshop on Cellular Neural Networks and Their Applications,* Istanbul, Turkey, 2006, pp. 1–6.

O. Drbohlav and M. Chantler, "On optimal light configurations in photometric stereo", *Tenth IEEE International Conference on Computer Vision,* Beijing, China, 2005, pp. 1707–1712.

T. Duckett, A. Lilienthal, and C. Valgren, "Incremental topological mapping using omnidirectional vision", 2006 IEEE/RSJ International Conference on Intelligent Robots and Systems, October 2006, pp. 3441–3447 .

E. Dunn, G. Olague, E. Lutton, and M. Schoenauer, "Pareto optimal sensing strategies for an active vision system", *IEEE Congress on Evolutionary Computation,* Portland, USA, 2004, pp. 457–463.

E. Dyllong and A. Kreuder, "Optimal reconstruction of mode shapes using nonuniform strain sensor spacing", *1999 IEEE/ASME International Conference on Advanced Intelligent Mechatronics*, 1999, pp. 150–155.

H. D. Einhorn, "Unified glare rating (UGR): Merits and application to multiple sources", Lighting Research Technology, Vol. 30, No. 2, 1998, pp. 89–93.

P. Eisert, E. Steinbach, and B. Girod, "Automatic reconstruction of stationary 3-D objects from multiple uncalibrated camera views", *IEEE Transactions on Circuits and Systems for Video Technology*, Vol. 10, No. 2, pp. 261–277, March 2000.

K. El Sahmarani, Z. Simeu-Abazi, and P. Ladret, "Vision system for command and fault detection", *International Conference on Service Systems and Service Management,* Troyes, France, 2006, pp. 1648–1652.

A. Elnagar, "Optimal error discretization under depth and range constraints", *Pattern Recognition Letters*, Vol. 19, No. 9, pp. 879–888, July 1998.

B. Elson and P. K. Khosla, "Integrating sensor placement and visual tracking strategies", *1994 IEEE International Conference on Robotics and Automation*, May 1994, Vol. 2, pp. 1351–1356.

T. Eltoft and R. J. P. deFigueiredo, "Illumination control as a means of enhancing image features in active vision systems", *IEEE Transactions on Image Processing*, Vol. 4, No. 11, pp. 1520–1530, November 1995.

S. B. Eng, E. Burdet, C. L. Teo, and J. E. Colgate, "Investigation of motion guidance with scooter cobot and collaborative learning", *IEEE Transactions on Robotics, [see also IEEE Transactions on Robotics and Automation]*, Vol. 23, No. 2, pp. 245–255, 2007.

R. M. Eustice, H. Singh, and J. J. Leonard, "Exactly sparse delayed-state filters for view-based SLAM", *IEEE Transactions on Robotics, [see also IEEE Transactions on Robotics and Automation]*, Vol. 22, No. 6, pp. 1100–1114, 2006.

F, G

S. Fangwu, L. Toma, W. Neddermeyer, and Z. Jianwei, "Precise online camera calibration in a robot navigating vision system", *IEEE International Conference on Mechatronics and Automation,* Niagara Falls, Canada, 2005, pp. 1277–1282.

A. A. Farag and A. E. Abdel-Hakim, "Image content-based active sensor planning for a mobile trinocular active vision system", *International Conference on Image Processing,* Singapore, 2004, pp. 2913–2916.

A. A. Farag and A. E. Abdel-Hakim, "Virtual forces for camera planning in smart vision systems,", *Seventh IEEE Workshops on Application of Computer Vision,* Breckenridge, CO, 2005, pp. 269–274.

A. A. Farag and A. E. Abdel-Hakim, "Image content-based active sensor planning for a mobile trinocular active vision system", *Proceedings of IEEE International Conference on Image Processing (ICIP'2004),* Singapore, October 24–27, 2004, Vol. II, pp. 193–196.

G. Farin, Curves and Surfaces for CAGD: A Practical Guide, 5th edition, San Francisco, CA : Morgan Kaufmann ; San Diego : Academic Press, 2002.

S. Fernand and Y. Wang, "Part I: Modeling image curves using invariant 3D object curve models: A path to 3D recognition and shape estimation from image contours", *IEEE Transactions on pattern analysis and machine intelligence*, Vol. 16, No. 1, pp. 1–12, 1994.

A. Ferreira, C. Cassier, and S. Hirai, "Automatic microassembly system assisted by vision servoing and virtual reality", *IEEE/ASME Transactions on, Mechatronics,* Vol. 9, No. 2, pp. 321–333, 2004.

M. Fiala, "Structure from motion using SIFT features and the PH transform with panoramic imagery", *The 2nd Canadian Conference on Computer and Robot Vision,* Victo, Canadian, 2005, pp. 506–513.

M. Fichtner and A. Grobmann, "A probabilistic visual sensor model for mobile robot localisation in structured environments", *IEEE/RSJ International Conference on Intelligent Robots and Systems,* Sendai, Japan, 2004, pp. 1890–1895.

M. A. Fischler and R. C. Bolles, "Random sample consensus: A paradigm for model fitting with applications to image analysis and automated cartography", *Graphics and Image Processing*, Vol. 24, No. 6, pp. 381–395, June 1981.

A. W. Fitzgibbon and A. Zisserman, "Automatic 3D model acquisition and generation of new images from video sequences", *European Signal Processing Conference*, Rhodes, Greece, 1998, pp. 1261–1269.

D. Fofi, J. Salvi, and E. Mouaddib, "Uncalibrated vision based on structured light", *IEEE International Conference on Robotics and Automation*, Seoul, Korea, May 2001.

J. S. Franco and E. Boyer, "Fusion of multiview silhouette cues using a space occupancy grid", *Tenth IEEE International Conference on Computer Vision,* Beijing, China, 2005, pp. 1747–1753.

W. J. Freeman, "Consciousness, intentionality, and causality", *Journal of Consciousness Studies*, Vol. 6, No. 11, pp. 143–172, November/December, 1999a.

W. J. Freeman, "Comparison of brain models for active vs. passive perception", *Information Sciences*, Vol. 116, No. 2, pp. 97–107, 1999b.

E. W. Frew, J. Langelaan, and J. Sungmoon, "Adaptive receding horizon control for vision-based navigation of small unmanned aircraft", 2006, p. 6.

P. Fua, "From multiple stereo views to multiple 3D surfaces", *International Journal of Computer Vision*, Vol. 24, No. 1, pp. 19–35, 1997.

P. Fua and G. Leclerc, "Taking advantage of image based and geometry-based constraints to recover 3-D surfaces", *Computer Vision and Image Understanding*, Vol. 64, No. 1, pp. 111–127, 1996.

A. Fusiello, "Uncalibrated euclidean reconstruction: A review", *Image and Vision Computing*, Vol. 18, No. 6–7, pp. 555–563, May 2000.

C. Georgiades, et al., "AQUA: An aquatic walking robot", *IEEE/RSJ International Conference on Intelligent Robots and Systems,* Sendai, Japan, 2004, pp. 3525–3531.

A. Georgiev and P. K. Allen, "Localization methods for a mobile robot in urban environments", *IEEE Transactions on Robotics, [see also IEEE Transactions on Robotics and Automation]*, Vol. 20, No. 5, pp. 851–864, 2004.

P. Geveaux, J. Miteran, S. Kohler, F. Truchetet, and E. Renier, "A lighting characterisation by a reliable method. Application to defect detection by artificial vision in industrial field", *Annual Conference IEEE Industrial Electronics Society*, Vol.3, pp. 1242–1245, September 1998.

G. L. Gimel'farb and R. M. Haralick, "Terrain Reconstruction from Multiple Views", *Proceedings of the 7th International Conference on Computer Analysis of Images and Patterns* (CAIP'97), Kiel, Germany, Springer-Verlag: Berlin, pp. 694–701, 1997.

C. Giraud and B. Jouvencel, "Sensor selection: A geometrical approach", *IEEE/SRJ International Conference on Intelligent Robots and Systems*, Pittsburgh, Pennsylvania, USA, August 5–9, 1995, pp. 555–560.

T. Goedeme, T. Tuytelaars, L. Van Gool, G. Vanacker, and M. Nuttin, "Feature based omnidirectional sparse visual path following", *IEEE/RJS International Conference on Intelligent Robots and Systems,* Edmonton, Canada, 2005, pp. 1806–1811.

W. H. Gray, C. Dumont, and M. A. Abidi, "Integration of multiple range and intensity image pairs using a volumetric method to create textured three-dimensional models", *Journal of Electronic Imaging*, Vol. 10, No. 1, pp. 252–262, January 2001.

G. Greger, P. Shirley, P. M. Hubbard, and D. P. Greenberg, "The irradiance volume", *IEEE Computer Graphics and Applications* , Vol. 18 , No. 2, pp. 32–43, 1998.

P. M. Griffin, L. S. Narasimhan, and S. R. Yee, "Generation of uniquely encoded light patterns for range data acquisition", *Pattern Recognition*, Vol. 25, No. 6, 1992, pp. 609–616.

E. Grosso and M. Tistarelli, "Active/dynamic stereo: a general framework", *IEEE Computer Society Conference on Computer Vision and Pattern Recognition*, 1993, pp. 732–4 .

X. G. Gu, M. M. Marefat, F. W. Ciarallo, "A robust approach for sensor placement in automated vision dimensional inspection", *1999 IEEE International Conference on Robotics and Automation*, May 1999, pp. 2602–2606.

J. K. Gudrun, A. S. Steven, and K. Takeo, "A physical approach to color image understanding", *International Journal of CV*, Vol. 4, 1990, pp. 7–38.

H. Gustafson, C. T. Lollini, et al., "Swarm technology for search and rescue through multi-sensor multi-viewpoint target identification", *The Thirty-Seventh Southeastern Symposium on System Theory,* Tuskegee, USA, 2005, pp. 352–356.

J. S. Gutmann, M. Fukuchi, and M. Fujita, "A floor and obstacle height map for 3D navigation of a humanoid robot," *2005 IEEE International Conference on Robotics and Automation,* Barcelona, Spain, 2005, pp. 1066–1071.

H

H. Hadj-Abdelkader, Y. Mezouar, and P. Martinet, "Path planning for image based control with omnidirectional cameras," *45th IEEE Conference on Decision and Control,* San Diego, USA, 2006, pp. 1764–1769.

S. Hadlington, "Breadth of vision", *IEE Review*, Vol. 51, No. 7, pp. 38–42, 2005.

P. Haigron, et al., "Depth-map-based scene analysis for active navigation in virtual angioscopy", *IEEE Transactions on Medical Imaging*, Vol. 23, No. 11, pp. 1380–1390, 2004.

T. Hamada, "Active vision", *International Neuroethology Congress*, Montreal, Canada, 1992.

M. Hamdi and A. Ferreira, "Microassembly planning using physical-based models in virtual environment"*, IEEE/RSJ International Conference on Intelligent Robots and Systems,* Sendai, Japan, 2004, pp. 3369–3374.

R. L. Hartley, E. Hayman, et al., "Camera calibration and the search for infinity", *Seventh IEEE International Conference on Computer Vision*, 1999, Vol. 1, pp. 510–517.

W. Hartmann, J. Zauner, M. Haller, T. Luckeneder, and W. Woess, "Shadow catcher": A vision based illumination condition sensor using ARToolKit, *IEEE International Augmented Reality Toolkit Workshop*, October 2003, pp.44–45.

N. Haruhiko, et al., "Multiple acoustical holography method for localization of objects in broad range using audible sound", *IEEE/RSJ International Conference on Intelligent Robots and Systems,* Beijing, China, 2006, pp. 1145–1150.

M. Harville and L. Dalong, "Fast, integrated person tracking and activity recognition with plan-view templates from a single stereo camera", *IEEE Computer Society Conference on Computer Vision and Pattern Recognition,* Washington, DC, USA, 2004, p. II-398.

P. Hebert, "A self-referenced hand-hold ranger sensor", *Proceedings of the 3rd International Conference on 3-D Digital Imaging and Modeling,* Ottawa, Canada, pp. 5–11, May 2001.

A. Henrichsen. *3D reconstruction and camera calibration from 2D images (Master of Science Thesis).* University of Cape Town, 2000.

D. C. Herath, K. Sarath, and D. Gamini, "Simultaneous localisation and mapping: A stereo vision based approach", *IEEE/RSJ International Conference on Intelligent Robots and Systems,* Beijing, China, 2006, pp. 922–927.

A. Heyden and K. Astrom, "Euclidean reconstruction from image sequences with varying and unknown focal length and principal point", *IEEE Computer Society Conference on Computer Vision and Pattern Recognition,* San Juan, Puerto Rico, pp. 438–443, 1997.

K. Higuchi, M. Hebert and K. Ikeuchi, "Building 3-D models from unregistered range images", *Graphical Models and Image Processing,* Vol. 57, No. 4, pp. 315–333, July 1995.

U. Hirofumi, K. T. Joo, and I. Seiji, "A color tracker employing a two-dimensional color histogram under unstable illumination," *SICE-ICASE International Joint Conference,* Busan, Korea, 2006, pp. 2725–2728.

Y. K. Ho and C. S. Chua, "3D model building", *Seventh International Conference on Image Processing and Its Applications,* 1999 , No. 1, pp. 270–274.

H. Hong, A. Narendra, and G. Chunyu, "Design analysis of a high-resolution panoramic camera using conventional imagers and a mirror pyramid," *IEEE Transactions on, Pattern Analysis and Machine Intelligence,* Vol. 29, No. 2, pp. 356–361, 2007.

Y. Hongshan, W. Yaonan, W. Qin, and K. Fei, "The design and development of active stereovision system for mobile robot navigation", *Fifth World Congress on Intelligent Control and Automation,* Hangzhou, China, 2004, pp. 3734–3737.

B. K. P. Horn, "Closed form solution of absolute orientation using unit quartenions", *Journal of Optical Society of America,* Vol. 4, No. 4, pp. 629–642, April 1987.

A. Hou, N. Qi, H. Zhang, and T. Liu, "Stereo mapping for a prototype lunar rover", 2006, pp. 1036–1041.

T. Huang and O. R. Mitchell, "Dynamic camera calibration", *1995 IEEE International Symposium on Computer Vision,* Coral Gables, USA, 1995, pp. 169–174.

W. Huakun, Z. Li, and Y. Feng, "Neural networks-based terrain acquisition of unmarked area for robot mowers", *ICARCV 2004 8th Control, Automation, Robotics and Vision Conference,* Kunming, China, 2004, pp. 735–740.

L. Huei-Yung, L. Jen-Hung, and W. Ming-Liang, "A visual positioning system for vehicle navigation", *IEEE International Conference on Intelligent Transportation Systems,* Vienna, 2005, pp. 534–539.

S. B. Hughes and M. Lewis, "Task-driven camera operations for robotic exploration", *IEEE Transactions on Systems, Man and Cybernetics, Part A,* Vol. 35, pp. 513–522, 2005.

X. Hui, R. Weibin, W. Jiatao, and C. Wei, "Optomechatronic design of integrated systems for microassembly of MEMS sensors", *2006 IEEE International Conference on Mechatronics and Automation,* Luoyang, China, 2006, pp. 859–864.

D. Q. Huynh, "Calibration of a structured light system: A projective approach", *IEEE Conference on Computer Vision and Pattern Recognition*, 1997, pp. 225–230.

I, J

K. Ikeuchi, J.-C. Robert, "Modeling sensor detectability with the VANTAGE geometric/sensor modeler", *IEEE Transactions Robotics and Automation*, Vol. 7, No. 6, pp.771–784, December 1991.

S. Inokuchi, K. Sato, et al., "Range-imaging system for 3-D object recognition", *Proceedings of 7th International Conference Pattern Recognition*, Montreal, Canada, pp. 806–808, 1984.

H. Ishida, H. Tanaka, H. Taniguchi, and T. Moriizumi, "Mobile robot navigation using vision and olfaction to search for a gas/odor source", *IEEE/RSJ International Conference on Intelligent Robots and Systems*, Sendai, Japan, 2004, pp. 313–318.

H. Ishida, T. Ushiku, S. Toyama, H. Taniguchi, and T. Moriizumi, "Mobile robot path planning using vision and olfaction to search for a gas source", *4th IEEE Conference on Sensors*, Irvine, USA, 2005, p. 4.

T. Iwata, M. Tokura, "Examination of the limitations of predicted lare sensation vote (PGSV) as a glare index for a large source", Lighting Research and Technology, Vol. 30, No. 2, 1998, pp. 81–88.

E. Izquierdo and X. Feng, "Modeling arbitrary objects based on geometric surface conformity", *IEEE Transactions on Circuits and Systems for Video Technology*, Vol. 9, No. 2, pp. 336–352, March 1999.

S. P. Jae and J. C. Myung, "Path planning with uncalibrated stereo rig for image-based visual servoing under large pose discrepancy", *IEEE Transactions on Robotics and Automation*, Vol. 19, No. 2, pp. 250–258, 2003.

O. Jokinen, "Self-calibration of a light striping system by matching multiple 3-D profile maps", *IEEE Second International Conference on 3-D Digital Imaging and Modeling*, Ottawa, 1999, pp. 180–190.

E. Jose and M. D. Adams, "Multiple line-of-sight predicted observations with millimetre wave radar for outdoor SLAM", *8th Control, Automation, Robotics and Vision Conference*, Kunming, China, 2004, pp. 155–160.

F. B. Jose and G. Domingo, "Visual homing navigation with two landmarks: The balanced proportional triangulation method", *IEEE/RSJ International Conference on Intelligent Robots and Systems*, Beijing, China, 2006, pp. 2289–2295.

B. Joseph, G. Sebastien, and D. Erick, "Rover localization through 3D terrain registration in natural environments", *IEEE/RSJ International Conference on Intelligent Robots and Systems*, Beijing, China, 2006, pp. 4121–4126.

R. Joshi and A. C. Sanderson, "Minimal representation multisensor fusion using differential evolution", *IEEE Transactions on Systems, Man and Cybernetics, Part A*, Vol. 29, No. 1, pp. 63–76, January 1999.

X. Jun, X. Jizhong, and X. Ning, "Fuzzy logic system for miniature climbing robots", *8th Control, Automation, Robotics and Vision Conference*, Kunming, China, 2004, pp. 2094–2099.

L. Junchuan, Z. Yuru, W. Tianmiao, X. Hongguang, and T. Zengmin, "Neuromaster: a robot system for neurosurgery", *IEEE International Conference on Robotics and Automation*, Barcelona, Spain, 2004, pp. 824–828.

A. Jung-Yup, P. Ill-Woo, L. Jungho, and O. Jun-Ho, "Experiments of vision guided walking of humanoid robot, KHR-2", *5th IEEE-RAS International Conference on Humanoid Robots*, Tsukuba, Japan, 2005, pp. 135–140.

K

S. Kagami, Y. Takaoka, Y. Kida, K. Nishiwaki, and T. Kanade, "Online dense local 3D world reconstruction from stereo image sequences", *2005 IEEE/RJS International Conference on Intelligent Robots and Systems,* Edmonton, Canada, 2005, pp. 3858–3863.

B. Kalyan, A. Balasuriya, T. Ura, and S. Wijesoma, "Sonar and vision based navigation schemes for autonomous underwater vehicles", *8th Control, Automation, Robotics and Vision Conference,* Kunming, China, 2004, pp. 437–442.

A. Kaneshige, et al., "Real time path planning based on the potential method with 3-D diffusion process for an overhead traveling crane", *5th Asia Control Conference,* Melbourne, Australia, 2004, pp. 1715–1722.

A. Kaneshige, et al., "The development of an autonomous overhead traveling crane with real time path planning based on the potential method", *International Conference on Control and Automation,* Budapest, Hungary, 2005, pp. 1079–1084.

S. B. Kang, "Catadioptric self-calibration", *2000 IEEE Conference on Computer Vision and Pattern Recognition,* Vol. 1, pp. 201–207, 2000.

F. Kececi, M. Tonko, H. -H. Nagel, and V. Gengenbach, "Improving visually servoed disassembly operations by automatic camera placement", *1998 IEEE International Conference on Robotics and Automation,* Leuven, Belgium, 16–20 May 1998, Vol. 4, pp. 2947–2952.

W. Kejun, L. Zhanying, and C. Hongxia, "Dynamic path tracking for visual guided vehicle based on vision information", *Sixth International Conference on Intelligent Systems Design and Applications,* Jinan, China, 2006, pp. 632–636.

J. M. Kelsey, J. Byrne, M. Cosgrove, S. Seereeram, and R. K. Mehra, "Vision-based relative pose estimation for autonomous rendezvous and docking", *2006 IEEE Aerospace Conference,* Piscataway, USA, 2006, p. 20.

R. Kelly, R. Carelli, et al., "Stable visual servoing of camera-in-hand robotic systems", *IEEE/ASME Transactions on Mechatronics,* Vol. 5, No. 1, pp. 39–48, March 2000.

T. E. Kim, S. H. Ryu, J. S. Choi, "Determining shape and reflectance properties objects by using diffuse illumination", *IEEE Asia Pacific Conference on Circuits and Systems,* 1996, pp. 520–523.

T. E. Kim and J. S. Choi, "3-D shape recovery of hybrid reflectance surface using indirect diffuse illumination", *1995 International Conference on Image Processing,* October 1995, Vol. 2, pp. 354–357.

W. S. Kim, "Computer vision assisted virtual reality calibration", *IEEE Transactions on Robotics and automation,* Vol. 15, No. 3, pp. 450–464, June 1999.

W. S. Kim, A. I. Ansar, R. D. Steele, and R. C. Steinke, "Performance analysis and validation of a stereo vision system", *2005 IEEE International Conference on Systems, Man and Cybernetics,* Hawaii, USA, 2005, pp. 1409–1416.

S. K. Kim, J. K. Paik, "Out-of-focus blur estimation and restoration for digital auto-focusing system", *Electronics Letters,* Vol. 34, No. 12, pp. 1217–1219, 1998.

T. Kitamura and D. Nishino, "Training of a leaning agent for Navigation-inspired by brain-Machine interface", *IEEE Transactions on Systems, Man and Cybernetics, Part B,* Vol. 36, No. 2, pp. 353–365, 2006.

H. Kobayashi, Y. Ohyama, H. Hashimoto, and S. Jin-Hua, "Manipulation of human behavior by distorted dynamics vision", *2006 SICE-ICASE International Joint Conference,* Busan, Korea, 2006, pp. 4446–4450.

S. Kovacic, A. Leonardis, and F. Pernus, "Planning sequences of views for 3-D object recognition and pose determination", *Pattern Recognition,* Vol. 31, No. 10, pp. 1407–1417, October 1998.

S. Kristensen, "Sensor planning with Bayesian decision theory", *Robotics and Autonomous Systems*, Vol 19, No. 3–4, pp. 273–286, March 1997.

B. Krose, R. Bunschoten, S. T. Hagen, B. Terwijn, and N. Vlassis, "Household robots look and learn: Environment modeling and localization from an omnidirectional vision system", *Robotics & Automation Magazine, IEEE*, Vol. 11, No. 4, pp. 45–52, 2004.

W. C. Krumbein, "The 'sorting out' of geological variables illustrated by regression analysis of factors controlling beach firmness", *Journal of Sedimentary Petrology*, Vol. 29, pp. 575–587, 1959.

A. Krupa and F. Chaumette, "Control of an ultrasound probe by adaptive visual servoing", *2005 IEEE/RJS International Conference on Intelligent Robots and Systems,* Edmonton, Canada, 2005, pp. 2681–2686.

R. Kumar and A. R. Hanson, "Robust methods for estimating pose and a sensitivity analysis", *Computer Vision Graphics and Image Processing*, Vol. 60, No. 3, pp. 313–342, November 1994.

A. Kurisu, Y. Yokokohji, and Y. Oosato, "Development of a laser range finder for 3D map-building in rubble"*, IEEE International Conference on Mechatronics and Automation,* Niagara Falls, Canada, 2005, pp. 1842–1847.

K. N. Kutulakos, C. R. Dyer, and V. J. Lumelsky, "Provable strategies for vision-guided exploration in three dimensions", *1994 IEEE International Conference Robotics and Automation,* 1994, pp. 1365–1372.

K. N. Kutulakos and C. R. Dyer, "Global surface reconstruction by purposive control of observer motion", *1994 IEEE Computer Society Conference on Computer Vision and Pattern Recognition*, 1994, pp. 331–338.

L

R. L. Lagendijk, J. Biemond, and D. E. Boekee, "Hierarchical blur identification", *Proceedings of the International Conference on Acoustics, Speech, and Signal Processing,* (ICASSP-90), Vol. 4, pp. 1889–1892, 1990.

D. Lamb, D. Baird, and M. Greenspan, An automation system for industrial 3D laser digitizing. Proceedings of the 2nd International Conference on 3D Digital Imaging and Modeling, Ottawa, Canada, October 1999, 148–157.

A. Lammen and A. Ruckelshausen, "ROBOLEO – An interactive seeing robot"*, 2004 IEEE Conference on Robotics, Automation and Mechatronics,* Singapore, 2004, pp. 1038–1041.

A. Laurentini, "Surface reconstruction accuracy for active volume intersection", *Pattern Recognition Letters*, Vol. 17, No. 12, pp. 1285–1292, 25 October 1996.

S. M. LaValle, *Planning Algorithms*, Cambridge University Press, Cambridge, UK, 2006.

G. Lee, J. Mou, and Y. Shen, "Sampling strategy design for dimensional measurement of geometric feature using coordinate measuring machine", *International Journal of Machine Tools and Manufacture*, Vol. 37, No. 7, pp. 917–934, 1997.

S. C. Lee and Y. Wang, "A general algorithm for recognizing small, vague, and imager-alike objects in a nonuniformly illuminated medical diagnostic image", *32nd Asilomar Conference on Signals, Systems & Computers*, 1998, No. 2, pp. 941–943.

B. Leibe, T. Starner, et al., "The perceptive workbench: Toward spontaneous and natural interaction in semi-immersive virtual environments", *2000 IEEE on Virtual Reality*, 2000, pp. 13–20.

P. Lehel, E. E. Hernayed, and A. A. Farag, "Sensor planning for a trinocular active vision systexn", *1999 IEEE Computer Society Conference on Computer Vision and Pattern Recognition*, 1999, Vol 2, pp. 306–312.

C. Leung and A. Al-Jumaily, "A hybrid system for multi-agent exploration"*, 2004 IEEE International Conference on Fuzzy Systems,* Budapest, Hungary, 2004, pp. 209–213.

I. Lewin and O. John, "Luminaire photometry using video camera techniques", *Journal of the Illuminating Engineering Society*, Vol. 28, No. 1, 1999, pp. 57–63.

Y. F. Li and S. Y. Chen, Automatic recalibration of an active structured light vision system, *IEEE Transactions on Robotics and Automation*, Vol. 19, No. 2, pp. 259–268, April 2003.

Y. F. Li and Z. Liu, "Uncertainty-Driven Viewpoint Planning for 3D Object Measurements", *Proceedings of the IEEE International Conference on Robotics and Automation*, Taipei, Taiwan, September 2003.

Y. F. Li, R. S. Lu, and S. Y. Chen, "Self-calibration of an active vision system for 3D robot vision", *Mechatronics and Machine Vision in Practice*, J. Billingley (ed), Research Study Press, Baldock, Hertfordshire, UK, September 2002, pp. 3–11.

G. Lidoris, K. Kuhnlenz, D. Wollherr, and M. Buss, "Information-based gaze direction planning algorithm for SLAM", *6th IEEE-RAS International Conference on Humanoid Robots*, Genova, Italy, 2006, pp. 302–307.

D. Liebowitz and A. Zisserman, "Combining scene and auto-calibration constraints", *7th International Conference on Computer Vision*, Kerkyra, Greece, pp. 293–300, September 1999.

C. Y. Lin, S. W. Shih, Y. P. Hung, and G. Y. Tang, "A new approach to automatic reconstruction of a 3-D world using active stereo vision", *Computer Vision and Image Understanding*, Vol. 85, No. 2, pp. 117–143, 2002.

S. Lin and S.W. Lee, "Estimation of diffuse and specular appearance", *Proceedings of the 7th IEEE International Conference on CV*, Vol. 2, 1999, pp. 855–860.

X. Lin, J. Zeng, and Q. Yao, "Optimal sensor planning with minimal cost for 3D object recognition using sparse structured light images", *IEEE International Conference on Robotics and Automation*, Minneapolis, MN, USA, Vol. 4, 1996, pp. 3484–3489.

H. Q. Liu and X. Y. Lin, "Model-based next view planning by using rules – automatic feature prediction and detection", *1994 IEEE Computer Society Conference on Computer Vision and Pattern Recognition*, pp. 773–777.

L. Liu, R. Xing, K. Wang, and W. Wang, "Key techniques in robocup middle-sized soccer research", *2004 IEEE International Conference on Robotics and Biomimetics*, Kunming, China, 2004, pp. 529–534.

Y. Liu, T. S. Huang, and O. D. Faugeras, "Determination of camera location from 2-D to 3-D line and point correspondences", *IEEE Transactions on Pattern Analysis and Machine Intelligence*, Vol. 12, No. 1, pp. 28–37, January 1990.

H. C. Longuet-Higgins, "A computer algorithm for reconstructing a scene from two projections", *Nature*, Vol. 293, pp. 133–135, 1981.

A. Loutfi, S. Coradeschi, L. Karlsson, and M. Broxvall, "Putting olfaction into action: Using an electronic nose on a multi-sensing mobile robot", *2004 IEEE/RSJ International Conference on Intelligent Robots and Systems*, Sendai, Japan, 2004, pp. 337–342.

L. Lucchese, G. Doretto, and G. M. Cortelazzo, "Frequency domain estimation of 3-D rigid motion based on range and intensity data", *International Conference on Recent Advances in 3-D Digital Imaging and Modeling*, Ottawa, Canada, May 1997, pp. 107–112.

B. Ludington, E. Johnson, and G. Vachtsevanos, "Augmenting UAV autonomy", *Robotics & Automation Magazine, IEEE*, Vol. 13, No. 3, pp. 63–71, 2006.

D. P. K. Lun, T. C. Hsung, and D. D. Feng, "Efficient blind blur identification using discrete periodic Radon transform", *Proceedings of 2001 International Symposium on Intelligent Multimedia, Video and Speech Processing*, pp. 79–82, 2001.

C. Luo, T. Y. Hsu, and K. L. Su, "The development of a multisensor based intelligent security robot: Chung Cheng #1", *2005 ICM IEEE International Conference on Mechatronics,* Taipei, Taiwan, 2005, pp. 970–975.

C. Luo, P. K. Wang, T. Y. Hsu, and T. Y. Lin, "Navigation and mobile security system of intelligent security robot", *2005 IEEE International Conference on Industrial Technology,* Hong Kong, 2005, pp. 260–265.

M

Y. Ma , R. Vidal, J. Kosecka, and S. Sastry, "Kruppa's equations revisited: Its degeneracy, renormalization and relations to chirality", *European Conference on Computer Vision,* Trinity College Dublin, Ireland, 2000, pp. 561–577.

Y. Ma and R. Vidal, et al., "Camera self-calibration: Degeneracy resolution for Kruppa's equation", http://black1.csl.uiuc.edu/~yima/

D. Mackay, "Information based objective functions for Active Data Selection", Neural Computation, Vol. 4, No. 4, pp. 448–472, 1991.

B. Madison, "Improved target handoff for single cycle instrument placement", 2006, p. 13.

C. B. Madsen and H. I. Christensen, "A viewpoint planning strategy for determining true angles on polyhedral objects by camera alignment", *IEEE Transactions on pattern analysis and machine intelligence,* Vol. 19, No. 2, pp. 156–161, February 1997.

M. J. Magee, B. A. Boyter, C. H. Chien, and J. K. Aggrawal, "Experiments in intensity guided range sensing recognition of three-dimensional Objects", *IEEE Transactions on Pattern Analysis and Machine Intelligence,* Vol. 7, No. 6, pp. 629–637, August 1985.

S. S. Makhanov, S. Vannakrairoju, S. Kondo, "A new blur identification scheme", *Proceedings of the IEEE Region 10 Conference on TENCON 99,* Vol. 2, pp. 1315–1318, 1999.

E. Malis, "Visual servoing invariant to changes in camera-intrinsic parameters," *IEEE Transactions on Robotics and Automation,* Vol. 20, No. 1, pp. 72–81, 2004.

M. Manuel and B. Wolfram, "Multiple hypothesis tracking of clusters of people", *IEEE/RSJ International Conference on Intelligent Robots and Systems,* Beijing, China, 2006, pp. 692–697.

E. Marchand, "Control camera and light source positions using image gradient information", *2007 IEEE International Conference on Robotics and Automation,* Rome, Italy, 2007, pp. 417–422.

E. Marchand and F. Chaumette, "Active vision for complete scene reconstruction and exploration", *IEEE Transactions on Pattern Analysis and Machine Intelligence,* Vol. 21, No. 1, pp. 65–72 , January 1999.

L. Mariottini and D. Prattichizzo, "EGT for multiple view geometry and visual servoing: Robotics vision with pinhole and panoramic cameras", *Robotics & Automation Magazine, IEEE,* Vol. 12, No. 4, pp. 26–39, 2005.

J. F. Martin and L. Chiang, "Cost effective vision system for mobile robots", *22nd International Conference of the Chilean Computer Science Society,* Copiapo, Chile, 2002, pp. 43–51.

W. Martin, W. Daniel, H. Markus, and J. Zhang, "Multimodal people tracking and trajectory prediction based on learned generalized motion patterns," *2006 IEEE International Conference on Multisensor Fusion and Integration for Intelligent Systems,* Heidelberg, Germany, 2006, pp. 541–546.

J. R. Martinez-De-Dios and A. Ollero, "An illumination-robust robot infrared vision system for robotics outdoor applications", *Proceedings of 2004 World Automation Congress,* Vol. 15, pp. 413–418, 2004.

N. Massions and R. Fisher, "A best next view selection algorithm incorporation a quality criterion". *Proceedings of the Ninth British Machine Vision Conference*, Southampton, UK, September 1998, 780–789.

K. Masuda, S. Thompson, S. Kagami, and T. Kanade, "Verification of stereo vision based localization system", *2004 IEEE International Conference on Systems, Man and Cybernetics,* Hague, Netherlands, 2004, pp. 5435–5440.

G. Mather, "The use of image blur as a depth cue in human vision", University of Sussex, http://www.biols.susx.ac.uk/home/George_Mather/, 2000.

G. Mather and D. R. Smith, "Depth cue integration: Stereopsis and image blur", *Vision Research*, Vol. 40, pp. 3501–3506, 2000.

J. Maver and R. Bajcsy, "Occlusions as a guide for planning the next view", *IEEE Transactions on Pattern Analysis and Machine Intelligence*, Vol. 15, No. 5, pp. 417–433, May 1993.

J. Maver, A. Leonardis, and F. Solina, "Planning the next view using the max-min principle", *Computer analysis of images and patterns*, Vol. 15, pp. 543–547, September 1993.

R. McCluney, *Introduction to Radiometry and Photometry*, Boston, Artech House, 1994.

M. McIvor, "Nonlinear calibration of a laser stripe profiler", *Optical Engineering*, Vol. 41, No. 1, pp. 205–212, January 2002.

A. M. McIvor and R. J. Valkenburg, "Calibrating a structured light system", *Image & Vision Computing*, New Zealand, pp. 167–172, August 1995.

A. D. R. Mcquarrie and C. L. Tsai, *Regression and Time Series Model Selection*, Singapore, River Edge, N.J. : World Scientific, 1998.

J. Meguro, T. Hashizume, J. Takiguchi, and R. Kurosaki, "Development of an autonomous mobile surveillance system using a network-based RTK-GPS", *2005 IEEE International Conference on Robotics and Automation,* Barcelona, Spain, 2005, pp. 3096–3101.

J. Meguro, K. Ishikawa, et al., "Creating spatial temporal database by autonomous mobile surveillance system (a study of mobile robot surveillance system using spatial temporal GIS part 1)", *2005 IEEE International Workshop on Safety, Security and Rescue Robotics,* Kobe, Japan, 2005, pp. 143–150.

E. Menegatti, G. Cicirelli, et al., "Explicit knowledge distribution in an omnidirectional distributed vision system", *2004 IEEE/RSJ International Conference on Intelligent Robots and Systems,* Sendai, Japan, 2004, pp. 2743–2749.

E. Menegatti, A. Pretto, A. Scarpa, and E. Pagello, "Omnidirectional vision scan matching for robot localization in dynamic environments," *IEEE Transactions on Robotics,* Vol. 22, No. 3, pp. 523–535, 2006.

F. Michahelles, R. Wicki, and B. Schiele, "Less contact: Heart-rate detection without even touching the user", *Eighth International Symposium on Wearable Computers,* Arlington, USA, 2004, pp. 4–7.

G. Milighetti, T. Emter, B. Kuntze, D. Bechler, and K. Kroschel, "Combined visual-acoustic grasping for humanoid robots", *IEEE International Conference on Multisensor Fusion and Integration for Intelligent Systems,* Heidelberg, Germany, 2006, pp. 1–6.

A. Miller and N. Nguyen, "A fedorov exchange algorithm for d-optimal design", Applied Statistics, Vol. 43, No. 4, pp. 669–678, 1994.

Y. K. Min, C. Hyungsuck, and L. Hyunki, "An active trinocular vision system for sensing mobile robot navigation environments", *2004 IEEE/RSJ International Conference on Intelligent Robots and Systems,* Sendai, Japan, 2004, pp. 1698–1703.

Y. Mokri and M. Jamzad, "Omni-stereo vision system for an autonomous robot using neural networks", *18th Annual Canadian Conference on Electrical and Computer Engineering*, Saskatoon, Canada, 2005, pp. 1590–1593.

K. Morooka, H. Zha, and T. Hasegawa, "Computations on a spherical view space for efficient planning of viewpoints in 3-D object modeling", *Second International Conference on3-D Digital Imaging and Modeling*, 1999, pp. 138–147.

M. Muehlemann, "Lighting for Color-Based Machine Vision Inspection of Automotive Fuse Blocks, Illumination Technologies, Inc." (http://www.machinevisiononline.org), 2000.

R. Murrieta-Cid, L. Muoz, M. Alencastre, et al., "Maintaining visibility of a moving holonomic target at a fixed distance with a non-holonomic robot", *IEEE/RSJ International Conference on Intelligent Robots and Systems IROS 2005*, Edmonton Canada.

R. Murrieta-Cid, A. Sarmiento, S. Bhattacharya, and S. Hutchinson, "Maintaining visibility of a moving target at a fixed distance: The case of observer bounded speed", *In Proceedings of IEEE International Conference on Robotics and Automation ICRA 2004*.

S. A. Musman, P. E. Lehner, and C. Elsaesser "Sensor planning for elusive targets", *Mathematical and Computer Modelling*, Vol. 25, No. 3, pp. 103–115, February 1997.

N

M. Nashman, T. Hong, et al., "An integrated vision touch-probe system for dimension inspection tasks", SME Applied Machine Vision'96 Conference, Cincinnati, OH. 1996, Society of Manufacturing Engineers, pp. 243–255.

S. K. Nayar, K. Ikeuchi, and T. Kanade, "Surface reflection: physical and geometrical perspectives", IEEE Transactions on PAMI, Vol.13, No. 7, pp. 611–634, 1991.

N. Navab and O. D. Faugeras, "The critical sets of lines for camera displacement estimation: A mixed euclidean-projective and constructive approach", Technical report, INRIA, Sophia, Antipolis, July 1994.

L. E. Navarro-Serment, J. M. Dolan, and P. K. Khosla, "Optimal sensor placement for cooperative distributed vision", *IEEE International Conference on Robotics and Automation,* Barcelona, Spain, 2004, pp. 939–944.

B. J. Nelson, N. P. Papanikolopoulos, et al., "Robotic visual servoing and robotic assembly tasks", *IEEE Robotics and Automation Magazine*, Vol. 3, No. 2, pp. 23–31, June 1996.

R. Nera and A. T. Jorge, "Control of a fleet of vehicles using computer vision, cellular automaton and genetic trained behaviour", *2006 IEEE International Conference on Mechatronics,* Budapest, Hungary, 2006, pp. 483–487.

X. D. Nguyen, Y. Bum-Jae, et al., "Simple visual self-localization for indoor mobile robots using single video camera", *IEEE/RSJ International Conference on Intelligent Robots and Systems,* Sendai, Japan, 2004, pp. 3767–3772.

G. Novak and R. Springer, "An introduction to a vision system used for a MiroSOT robot soccer system", *Second IEEE International Conference on Computational Cybernetics,* Vienna, Austria, 2004, pp. 101–108.

O, P, Q

K. Ohno, T. Nomura, and S. Tadokoro, "Real-time robot trajectory estimation and 3D map construction using 3D camera", *IEEE/RSJ International Conference on Intelligent Robots and Systems,* Beijing, China, 2006, pp. 5279–5285.

K. Okada, T. Ogura, et al., "Humanoid motion generation system on HRP2-JSK for daily life environment", *IEEE International Conference on Mechatronics and Automation,* Niagara Falls, Canada, 2005, pp. 1772–1777.

J. J. Okamoto, M. Milanova, and U. Bueker, "Active perception system for recognition of 3D objects in image sequences", *5th International Workshop on Advanced Motion Control,* Coimbra., 1998, pp. 700–705.

B. Olaf, Z. Zoran, and K. Ben, "Sparse appearance based modeling for robot localization", *IEEE/RSJ International Conference on Intelligent Robots and Systems,* Beijing, China, 2006, pp. 1510–1515.

G. Olague and R. Mohr, "Optimal camera placement to obtain accurate 3D point positions", *Fourteenth International Conference on Pattern Recognition,* 1998, Vol. 1, pp. 8–10.

A. Ozawa, Y. Takaoka, Y. Kida, K. Nishiwaki, J. Chestnutt, J. Kuffner, J. Kagami, H. Mizoguch, and H. Inoue, "Using visual odometry to create 3D maps for online footstep planning", *2005 IEEE International Conference on Systems, Man and Cybernetics,* Hawaii, USA, 2005, pp. 2643–2648.

K. Pahlavan, et al., "Active vision as a methodology", *Active Perception,* Hillsdale, N.J. Lawrence Erlbaum Associates, 1993, pp. 19–46.

K. Panchapakesan, D. G. Sheppard, M. W. Marcellin, and B. R. Hunt, "Blur identification from vector quantizer encoder distortion", *IEEE Transactions on Image Processing,* Vol. 10, No. 3, pp. 465–469, March 2001.

D. Papadopoulos, "3-D Sensing", 2001, http://perso.club-internet.fr/dpo/ numerisation3d/

D. Papadopoulos-Orfanos and F. Schmitt, "Automatic 3-D digitization using a laser rangefinder with a small field of view", *International Conference on Recent Advances in 3-D Digital Imaging and Modeling,* 1997, pp. 60–67.

J. C. Paricio Fernandes and J. A. B. Campos Neves, "Angle invariance for distance measurements using a single camera", *2006 IEEE International Symposium on Industrial Electronics,* Montreal, Canada, 2006, pp. 676–680.

L. E. Parker, B. Kannan, T. Fang, and M. Bailey, "Tightly-coupled navigation assistance in heterogeneous multi-robot teams", *IEEE/RSJ International Conference on Intelligent Robots and Systems,* Sendai, Japan, 2004, pp. 1016–1022.

J. L. Paul, "Smart sensor Web: Tactical battlefield visualization using sensor fusion", *Aerospace and Electronic Systems Magazine, IEEE,* Vol. 21, No. 1, pp. 13–20, 2006.

R. Pito, "A solution to the next best view problem for automated surface acquisition", *IEEE Transactions on Pattern Analysis and Machine Intelligence,* Vol. 21, No 10, pp. 1016–1030, October 1999.

M. Pollefeys, R. Koch, and L. Van Gool, "Self-calibration and metric reconstruction in spite of varying and unkonwn internal camera parameters", *International Conference on Computer Vision,* Bombay, India, 1998, pp. 90–96.

D. Popa, M. Rakesh, M. Manoj, S. Jeongsik, and H. Stephanou, "M3-modular multi-scale assembly system for MEMS packaging", *IEEE/RSJ International Conference on Intelligent Robots and Systems,* Beijing, China, 2006, pp. 3712–3717.

H. Pottmann, S. Leopoldseder, and M. Hofer, "Simultaneous registration of multiple views of a 3D object", *Archives of the Photogrammetry, Remote Sensing and Spatial Information Sciences,* Vol. XXXIV, Part 3A, Commisiion III, pp. 265–270, 2002.

F. Prieto, T. Redarce, P. Boulanger, and R. Lepage, "CAD-based range sensor placement for optimum 3D data acquisition", *Second International Conference on 3-D Digital Imaging and Modeling,* 1999, pp. 128–137.

A. Pun, T. I. Alecu, G. Chanel, J. Kronegg, and S. Voloshynovskiy, "Brain-computer interaction research at the computer vision and multimedia laboratory, University of Geneva", *IEEE Transactions on Rehabilitation Engineering,* Vol. 14, No. 2, pp. 210–213, 2006.

F. Qiang and X. Cunxi, "A study on intelligent path following and control for vision-based automated guided vehicle", *Fifth World Congress on Intelligent Control and Automation,* Hangzhou, China, 2004, pp. 4811–4815.

Y.-F. Qu, Z.-B. Pu, Y.-A. Wang, and G.-D. Liu, "Design of self-adapting illumination in the vision measuring system", *International Conference on Machine Learning and Cybernetics,* Vol. 5, November 2003, pp. 2965–2969.

S. Quan, X. Ning, C. Heping, and C. Yifan, "Calibration of robotic area sensing system for dimensional measurement of automotive part surfaces", *2005 IEEE/RJS International Conference on Intelligent Robots and Systems,* Edmonton, Canada, 2005, pp. 1526–1531.

R

J. Racky and M. Pandit, "Active illumination for the segmentation of surface deformations", *International Conference on Image Processing,* 1999, pp. 41–45.

M. Rahman, et al., "Architecture of the vision system of a line following mobile robot operating in static environment", *9th International Multitopic Conference,* Karachi, Pakistan, 2005, pp. 1–8.

A. Razdan and M. Bae. "A hybrid approach to feature segmentation of triangle meshes" *Computer Aided Design,* Vol. 35, No. 9, pp. 783–789, August 2003.

M. K. Reed, "Solid model acquisition form range imagery". Ph.D. dissertation. Columbia University, New York, NY. 1998.

M. Reed and P. Allen, "Constraint-based sensor planning for scene modeling", *IEEE International Symposium on Computational Intelligence in Robotics and Automation,* 1999, pp. 131–136.

M. K. Reed, P. K. Allen, et al., "3-D modeling from range imagery: an incremental method with a planning component 3-D Digital Imaging and Modeling", *International Conference on Recent Advances in 3D Imaging and Modeling,* May 1997, pp. 76–83.

S. J. Reeves and R. M. Mersereau, "Blur identification by the method of generalized cross-validation", *IEEE Transactions on Image Processing,* Vol. 1, No. 3, pp. 301–311, 1992.

I. D. Reid, "Projective calibration of a laser-stripe range finder", *Image and Vision Computing,* Vol. 14, pp. 659–666, 1996.

B. Reitinger, C. Zach, and D. Schmalstieg, "Augmented reality scouting for interactive 3D reconstruction", *2007 IEEE Virtual Reality Conference,* Charlotte, USA, 2007, pp. 219–222.

P. Remagnino, J. Illingworth, J. Kittler, and J. Matas, "Intentional control of camera look direction and viewpoint in an active vision system", *Image and Vision Computing,* Vol. 13, No. 2, pp. 79–88, March 1995.

K. Rice, J. Le Moigne, and P. Jain, "Analyzing range maps data for future space robotics applications", 2006, p. 1.

R. Rimey and C. Brown, "Where to look next using a bayes net: Incorporating geometric relations", *Second European Conference on Computer Vision,* Santa Margherita Ligure, Italy, pp. 542–550, May 1992.

H. Rivera-Rios, S. Fai-Lung, and M. Marefat, "Stereo camera pose determination with error reduction and tolerance satisfaction for dimensional measurements", *IEEE International Conference on Mechatronics and Automation,* Niagara Falls, Canada, 2005, pp. 423–428.

R. Rodrigo and J. Samarabandu, "Monocular vision for robot navigation", *IEEE International Conference on Mechatronics and Automation,* Niagara Falls, Canada, 2005, pp. 707–712.

L. Ronen, R. Ehud, and S. Ilan, "Landmark selection for task-oriented navigation", *IEEE/RSJ International Conference on Intelligent Robots and Systems,* Beijing, China, 2006, pp. 2785–2791.

F. Rooms, "Estimating image blur in the wavelet domain", *Proceedings IEEE International Conference on Acoustics, Speech, and Signal Processing*, Vol. 4, pp. 4190–4190, 2002.

S. D. Roy, S. Chaudhury, S. Banerjee, "Isolated 3D object recognition through next view planning", *IEEE Transactions on Systems, Man and Cybernetics, Part A*, Vol. 30, No. 1, pp. 67–76, January 2000.

A. Rosenfeld, "Computer vision: Basic principles", *Proceedings of the IEEE*, 1988, pp. 863–868.

S

M. Saadatseresht, F. Samadzadegan, and A. Azizi, "ANN-based visibility prediction for camera placement in vision metrology", *First Canadian Conference on Computer and Robot Vision,* London, Canada, 2004, pp. 188–194.

Saeedi, P. D. Lawrence, and D. G. Lowe, "Vision-based 3-D trajectory tracking for unknown environments", *Robotics, IEEE Transactions on* Vol. 22, No. 1, pp. 119–136, 2006.

H. Saito and M. Kimura, "Shape modeling of multiple objects from shading images using genetic algorithms", *IEEE Internsational Conference on Systems, Man and Cybernetics*, 1996, Vol. 4, pp. 2463–2468.

J. Salvi, J. Pags, J. Batlle. "Pattern codification strategies in structured light systems", *Pattern Recognition*, Vol. 37, No. 4, pp. 827–849, April 2004.

G. Sansoni, M. Carocci, R. Rodella, "Calibration and performance evaluation of a 3-D imaging sensor based on the projection of structured light", *IEEE Transactions on Instrumentation and Measurement*, Vol. 49, No. 3, pp. 628–636, June 2000.

P. Santana, J. Barata, et al., "A multi-robot system for landmine detection", *10th IEEE Conference on Emerging Technologies and Factory Automation,* Catania, Italy, 2005, p. 8.

A. Sarmiento, et al., A Sample-based convex cover for rapidly finding an object in a 3-D environment, *IEEE International Conference on Robotics and Automation*, 2005, Barcelona, Spain.

I. Sato, Y. Sato, and K. Ikeuchi, "Illumination distribution from shadows", *IEEE Computer Society Conference on CVPR*, 1999, pp. 306–312.

A. E. Savakis and H. J. Trussell, "Blur identification by residual spectral matching", *IEEE Transactions on Image Processing*, Vol. 2, No. 2, pp. 141–151, April 1993.

R. Schneider and L. Kobbelt, "Geometric fairing of irregular meshes for free-form surface design". *Computer Aided Geometric Design*. Vol. 18, No.4, pp. 597–604, May 2001

W. R. Scott, G. Roth, and J. F. Rivest, View planning for automated three-dimensional object reconstruction and inspection. *ACM Computer Surveys*, Vol. 35, No. 1, pp. 64–96, March 2003.

Y. Seo and K. S. Hong, "Theory and practice on the self-calibration of a rotating and zooming camera from two views", *IEEE Proceedings on Vision, Image and Signal Processing*, Vol. 148, No. 3, pp. 166–172, June 2001.

V. Sequeira, J. G. M. Gonçalves, M. I. Ribeiro, "Active view selection for efficient 3D scene reconstruction", *13th International Conference on Pattern Recognition* (ICPR'96) - Track1, Vienna, Austria, pp. 815–819, Aug 1996.

V. Sequeira, K. Ng, E. Wolfart, J. G. M. Gonçalves, D. C. Hogg, "Automated reconstruction of 3D models from real environments", *Journal of Photogrammetry and Remote Sensing*, Vol. 54, pp. 1–22, Feb 1999.

V. Sequeira, J. G. M. Gonçalves, et al., "3D environment modelling using laser range sensing", *Robotics and Autonomous Systems*, Vol. 16, No. 1, pp. 81–91, Nov. 1995.

J. Shamir, *Optical systems and processes*, Bellingham, WA: Spie, 1999.

L. Shapiro, "A Domain-Model Approach to Reconstruction of 3D Environments for Virtual Reality", University of Washington, 1995, http://www.cs. washington.edu/homes/shapiro/

A. Shashua and S.Toelg, "The quadric reference surface: Theory and applications", *International Journal Computer Vision*, Vol. 23, No. 2, pp. 185–198, 1997.

W. Sheng, N. Xi, et al., Viewpoint reduction in vision sensor planning for dimensional inspection. *Proceedings of the 2003 IEEE International Conference on Robotics, Intelligent Systems and Signal Processing*, ChangSha, China, October 2003, 249–254.

F. Schramm and G. Morel, "A calibration free analytical solution to image points path planning that ensures visibility", *IEEE International Conference on Robotics and Automation,* Barcelona, Spain, 2004, pp. 485–490.

F. Schramm, A. Micaelli, and G. Morel, "Calibration free path planning for visual servoing yielding straight line behaviour both in image and work space", *2005 IEEE/RJS International Conference on Intelligent Robots and Systems,* Edmonton, Canada, 2005, pp. 2216–2221.

C. Schwarz and N. da Vitoria Lobo, "The camera-driven interactive table," *2007 IEEE Workshop on Applications of Computer Vision,* Austin, USA, 2007, p. 50.

L. Ser-Nam, et al., "Fast illumination-invariant background subtraction using two views: error analysis, sensor placement and applications", *IEEE Computer Society Conference on Computer Vision and Pattern Recognition,* San Diego, USA, 2005, pp. 1071–1078.

A. Shacklock, "Multiple-view multiple-scale navigation for micro-assembly", *IEEE International Conference on Robotics and Automation,* Barcelona, Spain, 2004, pp. 902–907.

T. Shen, J. Huang, and C. Menq, "Multiple-sensor integration for rapid and high-precision coordinate metrology", *IEEE/ASME Transaction Mechatronics*, Vol. 5, No. 2, pp. 110–121, 2000.

T. Shen, J. Huang, and C. Menq, "Multiple-sensor planning and information integration for automatic coordinate metrology", *Journal of Computing and Information Science in Engineering*, Vol. 1, pp. 167–179, 2001.

W. Sheng, Y. Shen, and N. Xi, "Mobile Sensor Navigation with Miniature Active Camera for Structure Inspection", *2006 IEEE/RSJ International Conference on Intelligent Robots and Systems*, October 2006, Beijing, pp. 1177–1182.

L. Shicai, T. Dalong, and L. Guangjun, "Vision-based formation control of mobile robots with relative motion states", *2005 IEEE International Conference on Robotics and Biomimetics,* Hong Kong, 2005, pp. 72–76.

G. Shwartz, "Estimating the Dimension of a Model", *Annals of Statistics*, Vol. 6, pp. 461–464, 1978.

M. Siddiqui and S. Sclaroff, "Surface reconstruction from multiple views using rational B-splines", *CVPR: Technical Sketches*, 2001.

F. Si-Yao, Z. Yun-Chu, C. Long, L. Zi-Ze, H. Zeng-Guang, and T. Min, "Motion based image deblur using recurrent neural network for power transmission line inspection robot", *International Joint Conference on Neural Networks,* Vancouver, Canada, 2006, pp. 3854–3859.

P. Sinha and E. Adelson, "Recovering reflectance and illumination in a world of painted polyhedra", *Fourth International Conference on Computer Vision*, April 1993, pp. 156–163.

P. Skrzypczynski, Uncertainty Models of Vision Sensors in Mobile Robot Positioning, *International Journal Applied Mathematics and Computer Science*, 2005, Vol. 15, No. 1, pp. 73–88.

P. Slusallek and H. P. Seidel, "Vision-an architecture for global illumination calculations", *IEEE Transactions on Visualization and Computer Graphics*, Vol. 1 No. 1, pp. 77–96, March 1995.

A. Sluzek and T. C. Seong, "A feasibility study on a novel method of visual obstacle detection", *2004 International Conference on Image Processing,* Singapore, 2004, pp. 2447–2450.

C. Solanki, W. E. Dixon, C. D. Crane, and S. Gupta, "Uncalibrated visual servo control of robot manipulators with uncertain kinematics", *2006 45th IEEE Conference on Decision and Control,* San Diego, USA, 2006, pp. 3855–3860.

F. Solomon and K. Ikeuchi, "An illumination planner for Lambertian polyhedral objects", *IEEE International Conference on Robotics and Automation*, 1995, Vol. 2, pp. 1719–1725.

C. Song and S. Kim, "Reverse engineering: autonomous digitization of free-form surfaces on a CNC coordinate measuring machine", International Journal of Machine Tools and Manufacturing, Vol. 37, No. 7, pp. 1041–1051, 1997.

G. Soucy, F. Callari, and F. Ferrie, "Uniform and complete surface coverage with a robot-mounted laser rangefinder". *Proceedings of the IEEE/RSJ International Conference on Intelligent Robots and Systems*, Victoria, British Columbia, Canada, October 1998, 1682–1688.

A. D. Spence and M. J. Chantler, "Optimal illumination for three-image photometric stereo using sensitivity analysis", *Vision, Image and Signal Processing, IEEE Proceedings*, Vol. 153, No. 2, pp. 149–159, 2006.

T. Stahs and F. Wahl, "Fast and versatile range data acquisition", *IEEE/RSJ International Conference Intelligent Robots and Systems*, Raleigh, NC, pp. 1169–1174, July 1992.

I. Stamos and P. K. Allen, "Interactive sensor planning", *IEEE Computer Society Conference on Computer Vision and Pattern Recognition*, June 1998, pp. 489–494.

A. State, G. Welch, and A. Ilie. An Interactive Camera Placement and Visibility Simulator for Image-Based VR Applications, IS&T/SPIE 18th Annual Symp. on Electronic Imaging Science and Technology, San Jose, CA, January 2006.

J. Stauder, "Point Light Source Estimation from Two Images and Its Limits", *International Journal of Computer Vision*, Vol. 36, No. 3, 2000, pp. 195–220.

P. Steinhaus, M. Walther, B. Giesler, and R. Dillmann, "3D global and mobile sensor data fusion for mobile platform navigation", *2004 IEEE International Conference on Robotics and Automation,* Barcelona, Spain, 2004, pp. 3325–3330.

A. Stemmer, G. Schreiber, K. Arbter, and A. bu-Schaffer, "Robust Assembly of Complex Shaped Planar Parts Using Vision and Force," *IEEE International Conference on Multisensor Fusion and Integration for Intelligent Systems,* Heidelberg, Germany, 2006, pp. 493–500.

A. Stemmer, A. bu-Schaffer, and G. Hirzinger, "An analytical method for the planning of robust assembly tasks of complex shaped planar parts", *2007 IEEE International Conference on Robotics and Automation,* Rome, Italy, 2007, pp. 317–323.

J. Subrahmonia, D. B. Cooper and D. Keren, "Practical reliable bayesian recognition of 2D and 3D objects using implicit polynomials and algebraic invariants", *IEEE Transactions on Pattern Analysis and Machine Intelligence*, Vol. 18, No. 5, pp. 505–519, 1996.

M. Sugiyama and H. Ogawa, "Subspace information criterion for model selection", *neural computation*, Vol. 13, No. 8, pp. 1863–1889, 2001.

I. Suguru and M. Jun, "3D indoor environment modeling by a mobile robot with omnidirectional stereo and laser range finder", *IEEE/RSJ International Conference on Intelligent Robots and Systems,* Beijing, China, 2006, pp. 3435–3440.

L. Sun, H. Xie, W. Rong, and L. Chen, "Vision based integrated system for automated anodic bonding of MEMS sensors", *6th International Conference on Electronic Packaging Technology,* Shenzhen, China, 2005, pp. 596–600.

K. P. Sung, K. Munsang, and L. Chong-won, "Mobile robot navigation based on direct depth and color-based environment modeling", *IEEE International Conference on Robotics and Automation,* Barcelona, Spain, 2004, pp. 4253–4258.

T

J. Tae-Seok, K. Morioka, and H. Hashimoto, "Distributed sensor network for multi-agent motion tracking in intelligent space", *2006 SICE-ICASE International Joint Conference,* Busan, Korea, 2006, pp. 3716–3721.

M. Takahiro and I. Hiroshi, "Behavior selection and environment recognition methods for humanoids based on sensor history", *IEEE/RSJ International Conference on Intelligent Robots and Systems,* Beijing, China, 2006, pp. 3468–3473.

S. Takahiro, O. Akihisa, and Y. Shin'ichi, "Operation direction to a mobile robot by projection lights", *IEEE Workshop on Advanced Robotics and Its Social Impacts,* Nagoya, Japan, 2005, pp. 160–165.

S. Takezawa, D. C. Herath, and G. Dissanayake, "SLAM in indoor environments with stereo vision", *2004 IEEE/RSJ International Conference on Intelligent Robots and Systems,* Sendai, Japan, 2004, pp. 1866–1871.

K. A. Tarabanis, P. K. Allen, and R. Y. Tsai, "A survey of sensor planning in computer vision", *IEEE Transactionss on Robotics and automation,* Vol. 11, No.1, pp. 86–104, 1995a.

K. A. Tarabanis and R. Y. Tsai, "Computing viewpoints that satisfy optical constraints", *IEEE Computer Society Conference on Computer Vision and Pattern Recognition,* pp. 152–158, 1991.

K. A. Tarabanis, R. Y. Tsai, and P. K. Allen, "The MVP sensor planning system for robotic vision tasks", *IEEE Transactions on Robotics and Automation,* Vol. 11, No.1, pp. 72–85, 1995b.

K. A. Tarahunis, R. Y. Tsai, and S. Abrams, "Planning viewpoints that simultaneously satisly several feature detectability constraints for robotic vision", *Fifth International Conference Advanced Robotics,* 1991, pp. 1410–1415.

G. H. Tarbox, "Planning for complete sensor coverage in inspection", *Computer Vision and Image Understanding,* Vol. 61, No. 1, pp. 84–111, January 1995.

C. J. Taylor and D. J. Kriegman, "Structure and motion from line segments in multiple images", *IEEE Transactions on Pattern Analysis and Machine Intelligence,* Vol. 17, No. 11, pp. 1021–1032, Nov. 1995.

F. Tel and B. Lantos, "Projective reconstruction for robot vision system", *2006 IEEE International Conference on Mechatronics,* Budapest, Hungary, 2006, pp. 357–362.

E. Theunissen, G. J. M. Koeners, R. M. Rademaker, R. D. Jinkins, and T. J. Etherington, "Terrain following and terrain avoidance with synthetic vision", *24th Digital Avionics Systems Conference,* Washington, DC, 2005, pp. 4–11.

Third Dimension Software Ltd, "3D machine vision", 1999, http://www.trinicom.com/usr/third/

K. Thitikamol and P. J. Keleher, "Active correlation tracking", *19th IEEE International Conference on Distributed Computing Systems,* 1999, pp. 324–331.

N. Thomas, F. Wladimir, and H. Frank, "Large view visual servoing of a mobile robot with a pan-tilt camera", *IEEE/RSJ International Conference on Intelligent Robots and Systems,* Beijing, China, 2006, pp. 3307–3312.

U. Thomas, S. Molkenstruck, R. Iser, and F. M. Wahl, "Multi sensor fusion in robot assembly using particle filters", *2007 IEEE International Conference on Robotics and Automation,* Rome, Italy, 2007, pp. 3837–3843.

S. Thompson and S. Kagami, "Humanoid robot localisation using stereo vision", *5th IEEE-RAS International Conference on Humanoid Robots,* Tsukuba, Japan, 2005, pp. 19–25.

W. Thompson and J. Owen, "Feature-based reverse engineering of mechanical parts", *IEEE Transactions Robotics and Automation*, Vol. 15, No. 1, pp. 57–66, 1999.

S. Thrun, C. Martin, et al., "A real-time expectation-maximization algorithm for acquiring multiplanar maps of indoor environments with mobile robots", *Robotics and Automation, IEEE Transactions on*, Vol. 20, No. 3, pp. 433–443, 2004.

P. Torr, "Bayesian model estimation and selection for epipolar geometry and generic manifold fitting", *International Journal Computer Vision*, Vol. 50, No. 1, pp. 35–61, 2002.

B. Tovar, R. Murrieta-Cid, and S. LaValle. "Distance-optimal navigation in an unknown environment without sensing distances", *IEEE Transactions on Robotics*, Vol. 23, No. 3, 2007, pp. 506–518.

B. Triggs and C. Laugier, "Automatic camera placement for robot vision tasks", *IEEE International Conference on Robotics and Automation*, Nagoya (J) Vol. 2, 1995, pp. 1732–1737.

E. Trucco, M. Umasuthan, A. M. Wallace, and V. Roberto, "Model-Based Planning of Optimal Sensor Placements for Inspection", *IEEE Transactions on Robotics and Automation*, Vol. 13, No. 2, pp. 182–194, 1997.

E. Trucco and A. Verri, *Introductory Techniques for 3-D Computer Vision*, Upper Saddle River, N.J.: Prentice Hall, 1998.

R. Y. Tsai, "An efficient and accurate calibration technique for 3D machine vision", *Proceedings of IEEE Conference CVPR 1986*, Miami, USA, pp. 364–74, 1986.

R. Y. Tsai, "An Efficient and accurate camera calibration technique for 3D machine vision", *IEEE Computer Vision and Patten Recognition*, 1986, pp. 364–374.

U, V, W

T. Ura, Y. Kurimoto, H. Kondo, Y. Nose, T. Sakamaki, and Y. Kuroda, "Observation behavior of an AUV for ship wreck investigation," *Proceedings of MTS/IEEE OCEANS,* Washington, DC, USA, 2005, pp. 2686–2691.

W. R. Uttal, *Computational Modeling of Vision: The Role of Combination*, Marcel Dekker Inc., New York, 1999, pp. 137–143.

J. G. Wang and Y. F. Li, "3D object modeling using a binocular vision system", *16th IEEE Instrumentation and Measurement Technology Conference* 1999, Vol 2, pp. 684–689.

Y. Wang, "Locator and sensor placement for automated coordinate checking fixtures", *Journal of Manufacturing Science and Engineering*, Vol. 121, pp. 709–719, 1999

R. C. Weast, ed. *CRC Handbook of chemistry and physics*, 66th ed., CRC press: Boca Raton, FL, 1985, pp. F-195.

G. Wei, K. Arbter, and G. Hirzinger, "Active self-calibration of robotic eyes and hand-eye relationships with model identification", *IEEE Transactions on Robotics and Automation*, Vol. 14, No. 1, pp. 158–166, February 1998.

L. Weiguo, J. Songmin, T. Abe, and K. Takase, "Localization of mobile robot based on ID tag and WEB camera", *2004 IEEE Conference on Robotics, Automation and Mechatronics,* Singapore, 2004, pp. 851–856.

S. Weihua, X. Ning, S. Mumin, and C. Yifan, "CAD-guided sensor planning for dimensional inspection in automotive manufacturing", *Mechatronics, IEEE/ASME Transactions on*, Vol. 8, No. 3, pp. 372–380, 2003.

S. Weihua, S. Yantao, and X. Ning, "Mobile sensor navigation with miniature active camera for structure inspection", *IEEE/RSJ International Conference on Intelligent Robots and Systems,* Beijing, China, 2006, pp. 1177–1182.

S. Weik, "Registration of 3-D partial surface models using luminance and depth information", *International Conference on Recent Advances in 3-D Digital Imaging and Modeling,* Ottawa, Canada, may 1997, pp. 93–100.

C. Weilong, J. E. Meng, and W. Shiqian, "Illumination compensation and normalization using logarithm and discrete cosine transform", *8th Control, Automation, Robotics and Vision Conference,* Kunming, China, 2004, pp. 380–385.

J. W. Weingarten, G. Gruener, and R. Siegwart, "A state-of-the-art 3D sensor for robot navigation", *IEEE/RSJ International Conference on Intelligent Robots and Systems,* Sendai, Japan, 2004, pp. 2155–2160.

D. Weir, M. Milroy, C. Bradley, and G. Vickers, "Wrap-around B-spline surface fitting to digitized data with application to reverse engineering", *Transactions of ASME, Journal of Manufacturing Science and Engineering,* Vol. 122, No. 2, pp. 323–330, 2000.

S. Wenxia and S. Jagath, "Corridor line detection for vision based indoor robot navigation," 2006, pp. 1988–1991.

P. Whaite and F. Ferrie, "Autonomous exploration: driven by uncertainty", *IEEE Transactions Pattern Analysis and Machine Intelligence,* Vol. 19, No. 3, pp. 193–205, 1997.

J. Williams and L. Won-Sook, "Interactive virtual simulation for multiple camera placement", *IEEE International Workshop on Haptic Audio Visual Environments and Their Applications,* Ottawa, Canada, 2006, pp. 124–129.

S. Winkelbach and F. M. Wahl, "Shape from 2D edge gradients", *Lecture Notes in Computer Science 2191,* Springer, 2001, pp. 377–384.

W. Wolovich, H. Albakri, and H. Yalcin, "The precise measurement of free-form surface", *Transactions of AMSE, Journal of Manufacturing Science and Engineering,* Vol. 124, No. 2, pp. 326–332, 2002.

A. K. C. Wong, L. Rong, and X. Liang, "Robot vision: model synthesis for 3D objects", *IEEE/RSJ International Conference on Intelligent Robots and Systems,* 1998, Vol. 3, pp. 1820–1827.

L. M. Wong, C. Dumont, and M. A. Abidi, "Next best view system in a 3D object modeling task", *IEEE International Symposium on Computational Intelligence in Robotics and Automation,* 1999, pp. 306–311.

T. P. Wong and R. Jarvis, "Real time obstacle detection and navigation planning for a humanoid robot in an indoor environment", *IEEE Conference on Robotics, Automation and Mechatronics,* Singapore, 2004, pp. 693–698.

T. Woo and R. Liang, "Dimensional measurement of surfaces and their sampling", *Computer-Aided Design,* Vol. 25, No. 4, pp. 233–239, 1993.

T. Woo, R. Liang, C. Heieh, and N. Lee, "Efficient sampling for surface measurements", *Journal of Manufacturing Systems,* Vol. 14, No. 5, pp. 345–354, 1995.

C. Wu, D. Wang, and R. Bajcsy, "Acquiring 3D spatial data of a real object", *Computer Vision, Graphics and Image Processing,* Vol. 28, pp. 126–133, 1984.

J. Wyatt, "Steps toward the development of a chronic retinal implant," *2006 International Workshop on Wearable and Implantable Body Sensor Networks,* Cambridge, MA, USA, 2006, p. 1.

R. J. Valkenburg and A. M. McIvor, "Accurate 3D measurement using a structured light system", *Image and Vision Computing,* Vol. 16, No. 2, pp. 99–110 February 1998.

X, Y, Z

X. Xiao, Y. Fang, F. He, and B. J. Ma, "A Study of intelligent searching mobile robots under unknown environment", *25th Chinese Control Conference,* Harbin, China, 2006, pp. 1609–1614.

H. Xinhan, L. Xiadong, and W. Min, "Development of a robotic microassembly system with multi-manipulator cooperation", *2006 IEEE International Conference on Mechatronics and Automation,* Luoyang, China, 2006, pp. 1197–1201.

Y. Xu; J. Zhang, "Abstracting human control strategy in projecting light source", *IEEE Transactions on Information Technology in Biomedicine*, Vol. 5, No. 1, 2001, pp. 27–32.

X. Xudong and L. Kin-Man, "An efficient illumination compensation scheme for face recognition", *8th Control, Automation, Robotics and Vision Conference,* Kunming, China, 2004, pp. 1240–1243.

Y. Yagi, K. Tsuji, and M. Yachida, "Evaluation of iconic memory-based ORP navigation", *2004 IEEE International Conference on Robotics and Automation,* Barcelona, Spain, 2004, pp. 4271–4276.

T. Yamaguchi, S. Kato, K. Watabe, T. Kunitachi, and H. Itoh, "Humanoid robot navigation by probabilistic multiple stereo matching", *2004 International Symposium on Micro-Nanomechatronics and Human Science,* Nagoya, Japan, 2004, pp. 359–364.

M. Yan, V. N. Jeffrey, and G. Jing, "Multi-robot aggregation strategies with limited communication", *IEEE/RSJ International Conference on Intelligent Robots and Systems,* Beijing, China, 2006, pp. 2691–2696.

Z. Yan, B. Yang, and C. Menq, "Uncertainty analysis and variation reduction of three dimensional coordinate metrology. Part I: geometric error decomposition", *International Journal of Machine Tools and Manufacture*, Vol. 39, No. 8, pp. 1199–1217, 1999.

B. Yang and C. Menq, "Compensation for form error of end-milled sculptured surface using discrete measurement data", *International Journal of Machine Tools and Manufacture*, Vol. 33, No. 5, pp. 725–740, 1993

C. C. Yang and F. W. Ciarallo, "Minimizing the probabilistic magnitude of active vision errors using genetic algorithm", *IEEE International Conference on Systems, Man, and Cybernetics*, 1997, Vol. 3, pp. 2713–2718.

H. Yang and G. Welch. "Illumination insensitive model-based 3D object tracking and texture refinement". *Third International Symposium on 3D Data Processing, Visualization and Transmission (3DPVT)*, Chapel Hill, NC USA, June 14–16, 2006.

H. Yang, M. Pollefeys, et al., "Differential camera tracking through linearizing the local appearance manifold", *IEEE Computer Society Conference on CVPR* 2007.

Y. Yao and P. Allen, "Computing robust viewpoints with multi-constraints using tree annealing", *IEEE International Conference on Systems, Man and Cybernetics, Intelligent Systems for the 21st Century*, Vol. 2, pp. 993–998, 1995.

Y. Ye and J. K. Tsotsos, "Sensor planning for 3D object search", *Computer Vision and Image Understanding*, Vol. 73, No. 2, pp. 145–168, Feb 1999.

X. Yi, O. I. Camps, "3D object depth recovery from highlights using active sensor and illumination control", *IEEE Computer Society Conference on Computer Vision and Pattern Recognition*, 23–25 June 1998, pp. 253–259.

H. Yifeng and K. Gupta, "An adaptive configuration-space and work-space based criterion for view planning", *2005 IEEE/RJS International Conference on Intelligent Robots and Systems,* Edmonton, Canada, 2005, pp. 3366–3371.

H. Young-Guk, et al., "Service-oriented integration of networked robots with ubiquitous sensors and devices using the semantic Web services technology", *2005 IEEE/RJS*

International Conference on Intelligent Robots and Systems, Edmonton, Canada, 2005, pp. 3947–3952.

Q. Yu-Fu, P. Zhao-Bang, W. Ya-Ai, and L. Guo-Dong, "Design of self-adapting illumination in the vision measuring system", *2003 International Conference on Machine Learning and Cybernetics,* Xi'an, China, 2003, pp. 2965–2969.

S. Yu, P. Raniga, and I. Mohamed, "A smart camera for multimodal human computer interaction", *IEEE Tenth International Symposium on Consumer Electronics,* St. Petersburg, Russia, 2006, pp. 1–6.

Z. Yu, K. Wang and R. G. Yang, "Next best view of range sensor", *IEEE 22nd International Conference on Industrial Electronics, Control, and Instrumentation,* Vol. 1, 1996, pp. 185–188.

X. Yuan, "A mechanism of automatic 3D object modeling", *IEEE Transactions Pattern Analysis and Machine Intelligence.* Vol. 17, No. 3, pp. 307–311, March 1995.

J. Yuan and S. L. Yu, "End-Effector position-orientation measurement", *IEEE Transactions on Robotics and Automation,* Vol. 15, No. 3, pp. 592–595, June 1999.

P. Zanne, G. Morel, and F. Plestan, "Robust 3D vision based control and planning", *2004 IEEE International Conference on Robotics and Automation,* Barcelona, Spain, 2004, pp. 4423–4428.

H. Zeng-Guang, T. Min, et al., "Neural network methods for the localization and navigation of mobile robots", *18th Annual Canadian Conference on Electrical and Computer Engineering,* Saskatoon, Canada, 2005, pp. 1057–1060.

C. Zezhi, P. Pe, J. McDermid, et al., "Two-stage visual localisation: landmark-based pose initialisation and model-based pose refinement", *2005 IEEE/RJS International Conference on Intelligent Robots and Systems,* Edmonton, Canada, 2005, pp. 150–156.

H. Zha, K. Morooka and T. Hasegawa, "Next best viewpoint (NBV) planning for active object modeling based on a learning-by-showing approach", *Lecture Notes in Computer Science 1352,* 1998, pp. II–185.

J. Zhang and Y. Xu, "Modeling human strategy in controlling light source", *IEEE International Conference on Robotics and Automation,* 1999, Vol. 4, pp. 3140–3145.

Z. Zhang, "Modeling geometric structure and illumination variation of a scene from real images", *Sixth International Conference on Computer Vision,* 1998, pp. 1041–1046.

J. Zhao-Hui and A. Goto, "Visual sensor based vibration control and end-effector control for flexible robot arms", *2005 IEEE International Conference on Industrial Technology,* Hong Kong, 2005, pp. 383–388.

J. Zhen, A. Balasuriya, and S. Challa, "Recent developments in vision based target tracking for autonomous vehicles navigation", *IEEE Intelligent Transportation Systems Conference,* Toronto, Canada, 2006, pp. 765–770.

C. Zhenhe and J. Samarabandu, "Using multiple view geometry within extended Kalman filter framework for simultaneous localization and map-building", *IEEE International Conference on Mechatronics and Automation,* Niagara Falls, Canada, 2005, pp. 695–700.

J. Y. Zheng and A. Murata, "Acquiring 3D object models from specular motion using circular lights illumination", *6th International Conference on Computer Vision,* pp. 1101–1108, 1998.

Z. Zhigang, et al., "Keeping smart, omnidirectional eyes on you (adaptive panoramic stereovision)", *Robotics & Automation Magazine, IEEE,* Vol. 11, No. 4, pp. 69–78, 2004.

H. Zhou and S. Sakane, "Sensor planning for mobile robot localization based on probabilistic inference using Bayesian network", *International Symposium on Assembly and Task Plannings* (ISATP), pp. 7–12, 2001.

H. Zhou and S. Sakane, "Learning Bayesian network structure from environment and sensor planning for mobile robot localization", *IEEE International Conference on Multisensor Fusion and Integration for Intelligent Systems,* 2003, pp. 76–81.

Y. Zhuang, S. Tang, L. Liu, and W. Wang, "Motion control system in a hybrid architecture for middle-size soccer robot", *8th Control, Automation, Robotics and Vision Conference,* Kunming, China, 2004, pp. 2205–2210.

P. Zingaretti and E. Frontoni, "Appearance based robotics", *Robotics & Automation Magazine, IEEE,* Vol. 13, No. 1, pp. 59–68, 2006.

A. Zomet, L. Wolf and A. Shashua, "Omni-rig: linear self-recalibration of a rig with varying internal and external parameters", *IEEE International Conference on Computer Vision,* 2001, Vol. 1, pp. 135–141.

Index